HUMAN ORIGINS

The Story of our Species

HUMAN ORIGINS

The Story of our Species

UNDER THE DIRECTION OF YVES COPPENS

NICOLAS BUCHET
PHILIPPE DAGNEAUX

HACHETTE Illustrated

-7 | -4.5

THE FIRST BIPEDS

PRECURSORS OF MAN

Contemporary with the last dinosaurs, the first true mammals appeared some 70 million years ago. After the great saurians vanished, mammals took over every vacant niche in the ecosystem. They evolved and diversified: this was the start of the great odyssey of which we remain part today.

THE MARCH OF TIME

Some 10 million years ago a phenomenal event took place: a line of primates gave birth to the first pre-humans. The incredible story of our species had begun.

Life on Earth is relatively recent in comparison with the age of the Universe. It needed aeons – the metamorphosis of 'time' itself – to develop and diversify. The amazing odyssey of humankind has its origins in this universal crucible, this infinite fusion from which emerged both vast galaxies and the most elementary particles.

Time initially underwent a cosmological period, beginning with the very birth of the Universe, some 15 billion years ago. It was then that there appeared protons, neutrons and electrons, the basic building-blocks of all matter. The first stars took shape, synthesising the essential chemical elements of the cosmos.

Ten billion years later, 'time' became geological. The Earth, in fact, was formed 4.6 billion years ago, only a few hundred million years after the Sun, to become the platform on which an extraordinary phenomenon would emerge: life. At the present moment, our planet is the only haven of life, as far as we know, in the entire Universe. Time then entered its biological stage. Scientists are still uncertain as to the precise conditions and mechanisms required for this event. They know, however, that single-celled bacteria were the first forms of life to develop in the oceans, around 3.8 billion years ago. Next, roughly 600 million years ago, in waters close to the shore, there arose algae, sponges, jellyfish and worms. This was the biological explosion of the Cambrian period.

Cartilaginous fishes appeared 470 million years ago; some, driven by a necessity we can only guess at, began to colonise the land, at the same time, curiously, as plants. The oldest fossils of these organisms able to exist in the Earth's atmosphere date to the Silurian period, 435 million years ago. The first mammals, evolved from reptiles, made their appearance 200 million years ago. Contemporaries of the dinosaurs, they were the ancestors of most animals we know today, like cows, horses, elephants, monkeys ... These mammals represented a true innovation in living creatures: the young were born already formed, and their mothers possessed mammary glands to provide them with milk. They also had a covering of hair. In the case of apes, their facial muscles rendered them more expressive, and, above all, their brains were more complex.

Mammals initiated evolutionary lines that colonised specialised ecological niches. Foremost among them were the primates – apes in the broadest sense. Remains of the earliest primate, *Purgatorius*, have been discovered in North America; resembling a small rodent, it lived alongside the last dinosaurs, some 65 million years ago. Primates are distinguished by a precise number of teeth (44), opposable thumb and big toe, and forward-looking eye-sockets which reinforce the animal's perception of colours and three-dimensional vision.

Fifty million years ago, these apes climbed up into the forest canopies; while swinging by their arms, they discovered the upright posture. On the ground, they practised an elementary method of walking on two legs, over short distances, and made use of 'tools'. Like present-day chimpanzees, they probed ant and termite nests with pieces of stick or crushed shells between two stones.

Finally we come to 'human time'. At around 10 million years ago, a handful of apes sowed the seeds of our race. How exactly this occurred we still do not know, but the primates divided into two great families: on the one side the pre-humans, on the other, gorillas and chimpanzees (simians). The pre-humans made a somewhat discreet, though decisive, entrance on to the world's stage. In the course of thousands upon thousands of years, they traversed the planet in every direction, adapting to all the demands of climate and altitude, equally at home in the most arid or luxuriant conditions. Their branch had made a clean break with that of the simians. The ability to walk upright was their first great innovation, complemented by a prodigious development of their intelligence. In a very short period by the scale of terrestrial phenomena, they achieved pre-eminence over all earthly life forms. This period is the 10 million enthralling years that we are about to explore.

OPPOSITE PAGE
After the disappearance of the dinosaurs 65 million years ago, only a few reptiles (left-hand photo) survived. The opportunistic primates seized the niches left vacant by the departed giants. Right: Tropical rain forest.

BELOW
Bonobos, known as 'pygmy chimpanzees' despite being similar in size to the ordinary species, are probably our nearest relatives. In fact, we share 99.9% of our genes with them. Additionally, certain aspects of their behaviour are identical to ours.

The bonobos

A few years ago, researchers turned their attention to one of two modern species of chimpanzee, the bonobos, or *Pan paniscus*.
They discovered that these animals were very close to us: much closer than we might imagine given the distant date at which their evolutionary line separated from ours.
The bonobos have what is technically known as a more 'gracile' body shape (slenderer and more supple) than their direct cousin, *Pan troglodytes*. Like them, they live in social groups, but are less aggressive than their relatives or indeed human beings. Some of their behavioural patterns are identical to ours. They give the appearance of sexual promiscuity, but this serves mainly to even out relationships between individuals within the group. Interestingly, they are the only animals which couple face to face, human-style. But the most surprising thing about them is their mode of locomotion. Though they spend more time in the trees than other large African apes, they travel about quite happily on two legs on the ground. Studies have revealed that the same hip and back muscles are employed for both tree-climbing and walking.
The bonobos, for all that, are not our ancestors, but our closest relatives. They share with humans characteristics inherited more than 10 million years ago from a common African ancestor.

EAST SIDE STORY

Since its origin, Earth has been subject to a series of titanic upheavals: climatic changes, continents shifting together or apart at the whim of subterranean forces. Forced to adapt, hominids evolved into the genus *Homo*.

Primates in general and apes in particular originated from the warm, humid tropical regions of the period, especially North America and Europe. The evolution of one group, the hominids, was linked to a major climatic crisis lasting 8 million years. Periods of global warming and cooling followed one another, modifying the living conditions of animals and plants. This vital series of events was to start a chain reaction in Africa which transformed the hominids into the genus *Homo*.

For basically astronomical reasons, the Earth was cooling down. In fact, alterations in its distance from the Sun and in the angle of its axis during orbit were producing worldwide climatic changes. Arid zones were appearing in the tropics. The Antarctic ice cap was growing, while the overall level of the oceans fell. Large-scale animal migrations were in progress, particularly between Europe and Africa. Scientists are able to confirm this by studying the ocean floor and the records held by sediments from that period.

At the same time, large-scale tectonic movements produced upheavals in the geography of the African continent, chiefly in the eastern region. The direct cause of these surface movements was continental drift. Starting 17 million years ago, there was a resurgence of the phenomenon known as rifting – a sudden uplifting, and subsequent sinking, of the Earth's crust, accompanied by earthquakes and volcanic eruptions.

After an initial mountain barrier was thrown up on the eastern lip of the gigantic 6,000-km (3,700-mile) African rift during the Burdigalian era (20 million years ago), the crisis period of the Tortonian (10 million years ago) saw the uplift of a second chain, this time to the west of the rift. The result was a considerable modification of the regional climate.

Formerly characterised by an immense luxuriance of plants and particularly trees, eastern Africa now became parched. The clouds laden with Atlantic rain were unable to pass the new mountain barrier, so that tropical forest gave way to arboreal savannah and finally steppe. This difference in climate and vegetation between the east and west of the continent was constantly accentuated over the course of the following 6 million years.

Such a long-term alteration in environmental conditions also generated an adaptive response among animal species – and more specifically our ancestors – which would otherwise have had no reason to take place. During the period that these tectonic events were under way, the great hominoid apes were in fact very widely scattered across Africa. The separation of the continent into two zones led to difficulty in finding food and consequent modifications in these creatures' stature and mode of locomotion.

The apes of the great western forests formed the group which would evolve into gorillas, chimps and bonobos. To survive, those of the eastern savannah had to adapt to their new, more arid surroundings: it was they who would originate lines of pre-humans and humans. This was the thesis Yves Coppens advanced during a congress in Rome in 1982, under the title 'East Side Story'. But Coppens had gone even further. In the 1970s, after analysing the stratigraphy of the lower Omo valley in south-west Ethiopia, he had been able to demonstrate that another arid peak corresponded directly with the emergence of the genus *Homo*, between 2 and 3 million years ago. He nicknamed this the '(H)Omo Event'.

OPPOSITE
The savannah that confronted the first hominids must have looked something like this.

ABOVE
The Great African Rift is clearly visible from space, revealing its immense length. This computer-generated image shows landmasses already emerged but still subject to continental drift and yet to acquire their present-day contours.

Either side of the Great Rift Valley, life followed differing paths. The great apes were divided by the rift into two groups. To the west, they became our closest cousins; to the east, a new creature was about to emerge: Man.

The three African lines

Humans, chimpanzees and gorillas appear to have developed from three lines existing side by side in Africa 8 million years ago. The three genera are cousins, the result of branches diverging from a common ancestor. Is there any way we can verify this theory? As it happens, DNA testing of the three races has yielded invaluable information in the form of a remarkable correspondence: the genetic resemblance between chimp and Man is 99.9%. Additionally, studies of their chromosomes have revealed that Man and chimps share some genes, while others are only held in common by chimps and gorillas. The hypothesis of the common ancestor has been well and truly confirmed. We should emphasise, however, that these studies reveal only major tendencies; scientists are looking primarily to palaeontological data (examination of skeletal remains) for more precise evidence of links. The existence of these common characteristics also suggests that humans and chimps must have interbred at some period; briefly, however, since the two lines ended in complete separation. According to some researchers, such hybridisation could also have taken place between chimps and gorillas.

TOP
For more than 8 million years, an enormous geological upheaval shook the foundations of Africa, thrusting up mountains and violently rearranging the landscape.

Excavation techniques

Chance discoveries like those made by prospectors are often the starting-point for organised excavations using varied techniques adapted to the location. For caves and rich sites, objects are retrieved following a very precise plan. A grid is laid out, dividing the area into squares with 1-m (39-in) sides. The field-workers note the three-dimensional co-ordinates of each find: the first two (*x* and *y*) indicate the northing and easting with reference to the grid lines, the third gives the subsurface depth of the find and hence the layer it occupied. The object's position is transferred to a plan, and each is washed, sorted, numbered and classified before it is ready for detailed investigation. This methodical attention to detail allows researchers to reconstruct the site in three dimensions and study its history. This is essential, as an excavation is tantamount, in effect, to destruction of a site, and any information lost is lost for ever. It is like reading a book and tearing out the pages as one goes along. But this method only works for sites where it is possible to install large amounts of equipment, often resulting in a project lasting months or even years. With other sites, where conditions are less favourable, excavations have necessarily to be carried out with less exactness. Frequently, time is of the essence. In the desert, for instance, the nature of a site does not permit lengthy exploration: everything that can be must be recovered before sandstorms and the advance of the dunes obliterate it for good. So-called rescue digs, carried out prior to the construction of a road or large building, also necessitate swift action; again, the idea is to save as much material and information as possible before the site disappears permanently under concrete.

ABOVE
The cliffs of the African Rift Valley exhibit an impressive series of strata 3,000 metres (nearly 10,000 ft) thick, providing us with vital information about our origins.

AFRICA:
THE CRADLE OF HUMANITY

The vast majority of pre-human fossils discovered up to the present day derive from Africa, providing us with the most complete sequential record of our evolution. So was mankind born in that continent?

OPPOSITE
The areas around Lake Turkana, formerly Lake Rudolph, are the most rewarding for research in palaeoanthropology. The discoveries made there confirm our African ancestry.

A large number of palaeoanthropologists are in agreement that, according to our current knowledge, Africa was the cradle of humanity: the 'Out of Africa' theory. Since the beginning of the twentieth century, the continent has been the scene of what amounts to a 'bone-rush'. Thousands of fossils – human and pre-human – have been unearthed there, notably in Ethiopia, Tanzania, Mozambique, Kenya and, more recently, Chad.

Many of these remains are very old, but so far only two more or less complete skeletons have been recovered. One belonged to Lucy, an Ethiopian australopithecine from 3.2 million years ago; the other is that of the Turkana Boy (from Lake Turkana in Kenya), an example of *Homo ergaster* and 1.6 million years old. Most of these remains lie to the east of the Great Rift, suggesting a very probable evolution of the human line starting from that area. Two discoveries made in 2000 and 2001 – Orrorin, from 6 million years ago, and Toumai, from 7 million years ago – lend weight to the 'Out of Africa' hypothesis, though demanding some modification of Coppens' *East Side Story*.

It is true that the Rift Valley is a unique area of the world. During the course of its exploration, strata of great age – several million years old – have been brought to light. Examination of their stratigraphy allows us to read them like a book. This has resulted in the recovery of animal and vegetable remains found nowhere else on the planet.

In other regions of the world, it is necessary to dig down thousands of metres to reach strata of equivalent age. Furthermore, zones like Europe, Asia or the Americas have not yielded human fossils as old as those from Africa. And entire areas have not been excavated because of their dense vegetation. In short, present-day theories of evolution are based on reasonably precise sequences of samples recovered in Africa, spanning a period from 7 million to under 500,000 years ago.

Sedimentation and the formation of fossils

For an organism to come down to us in fossilised form, all of a certain number of conditions have to be met. Once dead, the animal or plant must be swiftly buried by some geological phenomenon, such as mud slides or rapid sedimentation, either under water or on land. Further, this burial must take place in an anaerobic environment, that is, lacking oxygen, otherwise the hard tissues like bone will degenerate too quickly. Marine fossils enjoy all these conditions, while those on land are subject to the whims of the winds, water, vegetation ... The majority of hominid remains have therefore come to light along the shores of lakes or watercourses. Caves and crevasses are good preservation sites, provided they are protected from rain and runoff water. On the other hand, acid soil, as in tropical forest zones, helps to destroy this evidence of the past. Once the water has disappeared and the organic tissues have dissolved, the bones begin the process of mineralisation. This is a long, random process, affecting only a tiny proportion of buried organisms. Their discovery, millions of years later, is very often due to chance; palaeontologists initially happen upon them, then excavate the area for more. Erosion (by wind, ice, glaciers) and tectonic or volcanic upheavals can reveal ancient landforms concealing fossils. All of this explains the small number of fossils compared with the billions of billions of organisms that have existed on Earth since life began.

OPPOSITE
As a result of climate changes, trees on the African continent have declined in number.

ABOVE
A skull abandoned in the open air, like this example, will leave no clues for palaeontologists.

DOUBLE PAGE OVERLEAF
The landscape of the arid African savannah.

THE UPRIGHT APE

Pre-humans began to walk on two legs: a novel mode of locomotion known as bipedalism. Finding it increasingly easy, they eventually made it permanent.

OPPOSITE AND ABOVE
Long at home in the branches, the first pre-humans would abandon the security of the trees for a new environment: the savannah.

Eight million years ago, a female ape gave birth to a peculiar little creature. On reaching adulthood, it took in hand the destiny of its modest group of hominids, which now lived on the arboreal savannah. The spaces between trees had become of vital significance, and continual trekking between them had forced these apes to walk upright. On all fours in the long grass they could not sight landmarks or spot approaching predators.

When they stood up on their hind legs, they knew where they were, where they were going, and the nature of the immediate surroundings. An erect posture, combined with permanent bipedalism for the sake of survival, provided them with a new way of looking at the world. In turn this new vision impelled them towards their destiny.

Other apes, of course, adopted the same two-legged gait for identical reasons, as well as to intimidate rivals: male gorillas, for example. But this gait is maintained only for very brief periods. Similarly with chimpanzees: their posture is even more erect than that of their close cousins, but causes a certain amount of fatigue. Man is the only primate able to remain upright and motionless for hours on end. It even appears, according to some authorities, that this is Man's normal rest position (because of his upright skeleton it makes little demand on the muscles), whereas apes rest in a sitting position.

Further, all the great apes revert to quadrupedal gait when they need to accelerate, 'knuckle-walking' with their weight on the outer side of the middle phalanges. Man, however, pushes off alternately with either foot, as if to walk very swiftly, which also enables him to run continuously and over long distances. However, despite these observations, experts still remain divided on the exact development of bipedalism.

Our ancestors' great innovation was to make this bipedalism permanent, assisted by a modification of the skeleton already found in the

Travelling on all fours through long grass, pre-humans could not take their bearings. To see better, they stood up on their hind legs. This was the start of their eternal wanderings in search of food and new trees. Little by little, they became bipedal.

great apes. Over the course of time, their skulls attained a position at the top of the vertebral column and not on its forward-facing extension, as is the case with the majority of mammals. In other words, their heads were repositioned with respect to the occipital opening, leading in turn to a development of the brain in the rear portion of the skull. As their bodies became more erect,

Bipedalism: interpreting the clues

How, we may ask, can we tell from mere fossilised bones whether an animal was bipedal? In fact, numerous parts of the skeleton provide anatomists with vital information on the way our ancestors got about. The elbow and knee joints tell us about a creature's ability to perform certain movements characteristic of walking upright. The shape of the femur and the joint where it connects with the pelvis hold clues as to whether it could stand erect in a more or less stable manner over a reasonable period. The configuration and size of the humerus determine its ability to climb. In the orang-utan, for instance, the upper limbs are highly elongated compared to the rest of the body, indicating the animal is superbly adapted for life in the tree canopy. By examining the occipital opening (*foramen magnum*), we can learn much about the animal's posture. This orifice, through which the vertebral column passes to link with the brain, occupies a differing position depending on posture and gait. Only Man has a vertical spinal column and a *foramen* sited in the same plane so that the base of the skull forms a right angle with the spine. Finally, bones also retain the marks of muscle attachments or insertions. From the size and shape of these areas, experts can deduce the characteristics of the associated muscles, and hence their function. Man, for instance, has one of the best-developed sets of gluteal muscles of all mammals, which proves that 6 million years ago – when Orrorin lived – he was already standing and walking erect.

their neck muscles underwent corresponding shrinkage, relaxing their hold on the skull. Their spines assumed a very distinct curvature, while the pelvis became more vertical. Once standing upright, pre-humans used their hands for all sorts of tasks beside grasping, an ability that would reveal itself extremely useful in their later adventures.

The first hominoid known to have experimented with real bipedalism is Orrorin, discovered in Kenya. His legs were still bandy, like his cousins', and his muscles were not as yet adapted to this kind of exercise. His gait would have been swaying and hesitant; he would only have been able to walk for an hour or two at the most. His muscles probably hurt, but he was on the right road. More than 2 million years later, in what would become Ethiopia, Lucy was still staggering along, but rather less clumsily. Her knees would be only slightly bent and, notably, her femurs more erect. Finally, with *Homo habilis*, our probable direct ancestor, this two-legged gait came to look almost like our own.

OPPOSITE AND ABOVE
Discovered in Kenya, Orrorin is the first hominoid known to have tried walking on two legs.

OPPOSITE AND LEFT
At 1.5 m (nearly 5 ft) tall, Orrorin felt confident and strong, and was able to spot danger arriving.

BELOW
Orrorin's bones bear traces of bite marks, undoubtedly those of a sabre-toothed tiger. The first hominids were still easy prey.

ORRORIN: THE PRIMATE WHO STARTED A REVOLUTION

Six million years ago, while still displaying similarities with the australopithecines, this hominid had already acquired human characteristics, including the ability to walk upright.

The year 2000, marking the end of a century and a millennium, was a particularly happy one for palaeoanthropologists. A French and Kenyan expedition, led by Brigitte Senut from the Muséum national d'histoire naturelle, Paris, and Martin Pickford, of the Collège de France, discovered the remains of a new hominid of the Miocene period, 6 million years old. On four sites in the Lukeino Formation, Kenya, the team identified 13 fragments of bone and teeth from six different individuals. This new species is presently known as *Orrorin tugenensis*: the first term means 'earliest man' in the Tugen language, the second refers to the Tugen hills where the remains were brought to light.

This creature exhibits a set of characteristics partly associated with future humans and partly with later australopithecines. The neck of his femur and the size of the joint give definite indications that he was bipedal. This bone is also longer by a third than Lucy's, signifying a very tall individual for the period. He would have been around 1.5 m (nearly 5 ft) and weighed between 30 and 40 kg (66 and 88 lb).

The femur alone reveals that Orrorin walked erect in two different but adjacent environments: forest and open country. The attachments of the gluteal muscles, much more developed than in the great apes, add conclusive proof of this early bipedalism.

Certain features of Orrorin's teeth display chimpanzee-like features, but the thickness of enamel on the molars is closer to that found in modern Man, confirming that this trait is ancestral among hominids. However, the first bones of the fingers, long, slender and curving, suggest a definite agility and travel by 'brachiation' – swinging through the tree canopy. Similarly, finely furrowed canine teeth show affinity with the great apes, both of the Miocene period and the present day.

Despite these clues, Orrorin's discoverers still know nothing about his origins. For the moment, he is placed somewhere between the great apes known as *Samburupithecus*, who flourished 9.5 million years ago, and the australopithecines, who only arrived on the scene 2 million years later. Perhaps he stands at the final divergence between apes and Man. Future discoveries of even older hominids will doubtless throw fresh light on the matter.

The demonstrable ability of this new species to walk erect has pushed back the date at which bipedalism is thought to have developed. Before Orrorin, a number of scientists agreed that the australopithecines were the first to use this method of locomotion some 4 million years ago. Orrorin thus represents an extraordinary find, in the sense that it forces the experts to revise their scenario of human evolution and, specifically, their ideas on the adoption of bipedalism. Orrorin's discovery triggered a series of further excavations which ended in the discovery in Ethiopia of two extremely ancient hominids. Orrorin also breathes new life into Yves Coppens' *East Side Story* hypothesis.

The team who discovered Orrorin noticed that all the bones bore traces of carnivore bites, leading to the conclusion that some of his kind were either killed by predatory carnivores or eaten after death. Thanks to their size and ability to walk on two legs, our ancestors were beginning to leave their mark on their environment, but they still lacked the means to defend themselves against every danger lurking on the savannah.

OPPOSITE
At present, Toumai is the sole known representative of the race that existed immediately after the separation between chimps and Man.

TOUMAI:
FATHER OF MANKIND?

Toumai (*Sahelanthropus tchadensis*) is the oldest pre-human discovered to date. His presence in Chad proves that these creatures occupied a much wider territory than experts previously thought.

Toumai died 7 million years ago in an area of arboreal savannah surrounding Lake Chad which, at that time, was an inland sea covering over 400,000 km^2 (155,000 square miles). In 2001 a French and Chadian team from the University of Poitiers, led by Michel Brunet, unearthed the remains of this creature whose name means 'Hope of Life' in the Goran language. A somewhat deformed skull, fragments of jawbone and a few teeth sent a shock wave through the close-knit world of palaeontology. What on earth was such a hominid doing at this period, more than 2,000 km (1,240 miles) west of the Rift Valley? What did this mean for the *East Side Story* theory?

Sahelanthropus tchadensis – 'The Hominid of the Chadian Sahel' – displays pre-human characteristics, such as a relatively flat face, and an occipital opening proving he was definitely bipedal. As for his teeth, Michel Brunet lays stress on the fact that there are no gaps between them, and that the enamel is thicker than is found on those of great apes. There are also ancestral features, such as the small brain capacity: some 350 cm^3 (approx. 21 cubic inches) as against 1,350 cm^3 (82 cubic inches) for modern Man, closer in fact to that of present-day chimpanzees.

Disputes among experts are, unofficially anyway, among the tools of science, and the finding of Toumai was no exception to the rule. For Michel Brunet, doubt was impossible: 'The characteristics of the skull indicate incontrovertibly that it belongs to the human branch.' But, for Brigitte Senut, who discovered Orrorin, the skull appeared to be from an ancestor of the gorillas: 'The discovery of a pre-gorilla would be even more important, as no fossil of a direct ancestor of the great African apes has yet been found.'

On this very specific point there hangs, in fact, a whole new perception of the evolution of the genus *Homo*. Toumai is the only known survivor, for the moment, from the period immediately subsequent to the separation between chimpanzees and hominids. This indicates a point of division much older, by more than 7 million years, than that envisaged by numerous – but not all – authorities. What is more, the discovery proves that the hominids of this epoch were far more widely distributed than was suggested by Yves Coppens' hypothesis. Last, one of the theories attempting to explain Toumai's presence in these parts at this period postulates that he was a representative of one or several branches of pre-humans that spread in increasing numbers from East Africa, but the tentative conclusions of the Poitiers team are that Toumai may have belonged to a group closely related to that which produced Orrorin a million years later.

So there are some hard questions to resolve. Did the migration between the Rift and Chad really take place, and, if so, in which direction? Could Toumai have simply been born in Chad? How do we classify him precisely in the phylogenetic tree of pre-humans, or, even more problematic, among the ancestors of Man?

Naming fossils

Since Linnaeus, in the eighteenth century, every known animal or plant species has had a two-part Latin name denoting its genus and species. The name of the genus indicates membership of a group with broadly shared characteristics, as *Homo* for Man, *Australopithecus* for the 'ape from southern Africa' or *Sahelanthropus* for the 'hominid from the Sahel'. The species name, with pre-humans, often refers to the place of discovery: *A. africanus* comes from Africa, his ancestor *A. afarensis* having been brought to light in the Afar region, while Lucy's contemporary *A. bahrelghazali* lived next to the River of Gazelles, in Chad ... Humans are named after some particular characteristic of the species. *Homo habilis* was *skilful* (Latin: *habilis*) because he made tools; *Homo erectus* stood *erect*, walking on two legs exactly like us. In Latin, *sapiens* means *wise*, hence we are *Homo sapiens*! But palaeoanthropologists like to give their discoveries nicknames, simpler to use when discussing them and for the media to handle. So Lucy's nickname was inspired by the Beatles' song *Lucy in the Sky with Diamonds*. On occasions, as with the *Australopithecus* from Chad, the name is in memory of a friend or relative: in this case, Abel (*see p. 48*). Sometimes the place of discovery will suggest a name, as with Cro-Magnon or Tautavel Man. Finally, scientists in the field include locals who name finds in their own languages, as Toumai in Goran and Orrorin in Tugen.

RIGHT
At 2,000 km (1,240 miles) from the 'Cradle of Humanity', Toumai led a peaceful existence with neighbours like the placid *Chalicoterus*, an extinct ancestor of the giraffe.

-4.5 -2.5

AUSTRALOPITHECUS

AFRICAN PIONEERS

Toumai, then Orrorin, have probably disappeared. Two million years slip by. Drought extends across the African continent, denuding large regions of savannah. The first known successors of the hominids have had to adapt to the pressures of a constantly changing climate, while numerous animal and plant species have become extinct. Among the survivors of the climate changes, several groups of hominids spread from Chad to southern Africa. Collectively, they are known as *Australopithecus*.

A GIRL CALLED LUCY

The middle valley of the Awash in Ethiopia has preserved the remains of a female *Australopithecus afarensis*. Named Lucy by her French and American discoverers, she apparently drowned in the river.

ABOVE
Communication with other species of *Australopithecus* cannot always have been easy, living as they did in virtually different worlds.

OPPOSITE
Lucy owes her name to the Beatles' song *Lucy in the Sky with Diamonds*, a favourite in the camp of her French and American discoverers.

In 1924, at Taung in South Africa, an Australian anatomist named Raymond Dart discovered the skull of a child, which he baptised *A. africanus* – 'the South African monkey'. He drew attention to the fact that its traits were at the same time simian and human. This was the beginning of a long series of discoveries culminating, in 1974, in the exhumation of 52 bones in Ethiopia's Afar Triangle. These bones form 40% of the skeleton of another type of australopithecine known as *A. afarensis*, whose representative was nick-named Lucy by her French and American dis-coverers – the Beatles' song *Lucy in the Sky with Diamonds* was a favourite on the camp gramo-phone. The Ethiopians preferred to call her Birkinesh or 'person of worth'.

This was very apt, as Lucy's remains proved to be of immense interest. To start with, the treasure trove of bones was extraordinary: a femur and tibia, the upper limbs, the joints of the knees, elbows and shoulders, the ribs, spine, pelvis and sacrum, the lower jaw and part of the skull. These yielded a wealth of information, each part revealing something about Lucy's lifestyle. Further, analysis of the earth in which she had been found led to her remains being dated to between 3.2 and 3.1 million years ago.

Lucy measures little more than 1 m (39 in) and weighs around 20 kg (44 lb). She was about 20 – old for such a creature – and must already have borne children. Her small skull – with a cranial capacity of around 400 cm³ (24 cubic inches) – did not prevent her brain from out-performing Orrorin's. The upper portion of her body resembles that of the apes: arms longer than the legs, an almost non-existent waist, curved phalanges and a system of joints typically adapted to a semi-arboreal lifestyle. On the other hand, her locomotory apparatus reveals more human-like developments: a wide, flaring pelvis, lumbar lordosis (double curva-ture of the spine), a short, broadened sacrum and a necked femur – all indicative of the ability to walk upright.

Australopithecus probably couldn't swim.
But curiosity drove him to cross running water –
which is undoubtedly how Lucy met her end.

In the open savannah, Lucy moved with a slightly stooping, swaying gait. Her legs were bent, with her weight resting on the outside of her feet. Already she was travelling longer distances than her ancestors, though she was still happy to take refuge in the trees or look for

part of her food there. Her hands were capable of grasping stones and branches, but still unable to exercise a powerful grip like those of humans. Her teeth – thickly enamelled – and the attachments of powerful chewing muscles reveal that her diet was primarily vegetarian – bulbs, leaves, fruits, roots, etc., which often had thick skins. Because of her highly varied environment, she would also consume ants and termites as well as the flesh of birds, small dead mammals and the like – but scantily, and in an opportunistic fashion. This method of feeding was at the root of the australopithecines' success. It also shows how well they knew the environment, thanks to their ever-developing intelligence.

Lucy undoubtedly met her death by drowning, as her remains were discovered in traces of a former river. More than 3 million years after she vanished, the world was deeply moved when Yves Coppens, Donald Johanson and Maurice Taïeb brought to light these 52 bones.

PRECEDING DOUBLE PAGE AND ABOVE
Faced with a river, Lucy (*A. afarensis*) hesitates, less bold than her *anamensis* cousins. 'Should I go across and join them? Or stay on this side, all alone?'

OPPOSITE
Lucy finally decides to cross. But the strong currents prove fatal.

Somebody else's grandmother!

Despite what the public and the media have long believed and her worldwide fame, Lucy was not the 'grandmother' of humanity. She is likely, in fact, to have represented a branch that diverged from our line of descent, the members of which disappeared leaving no progeny. Compared with contemporary or later australopithecines, she actually displays numerous archaic features, like her knee, that are closer to those of the great apes.
Research appears to indicate that several species of hominid evolved simultaneously, each in a quite specific region of Africa. It is as if Nature had carried out a number of trials before deciding on which variation should become Man's forebear.

OPPOSITE
The self-confident *A. anamensis* adopts a different strategy to acquire food. His impact on the environment far outstrips that of his predecessors.

RIGHT
A. anamensis is more opportunist than other australopithecines, and eats more meat. He becomes a serious rival to hyenas, wild dogs and other scavengers.

ANAMENSIS

A CHANGE OF REGIME

A contemporary of *A. afarensis*, *A. anamensis*, another form of australopithecine, appears to be the prime candidate for the title of our true ancestor. He displays more numerous human characteristics, particularly in regard to bipedalism, than Lucy.

The australopithecines were not represented solely by *A. afarensis*. Indeed, at this period, at least five related species were extant, represented by over 300 individuals. Scattered across three regions of Africa, they lived between 4.5 and 2.5 million years ago. Skeletons and fragments of bones and teeth unearthed since the 1920s in Kenya, Tanzania, Ethiopia, southern Africa and Chad reveal a mosaic of features both pre-human and archaic, complicating the task of deciding which of them belonged to Man's real ancestor.

The first, *Australopithecus anamensis*, lived at the same time as Lucy, with the oldest specimen so far discovered dating to 4.5 million years ago. *Anam* is a Turkana word meaning 'lake', as the earliest find was on the shores of Lake Turkana in Kenya. Though he appeared in the same era as *A. afarensis*, he presents certain anatomical features closer to Man. In fact, examination of the tibia and the knee, which carried his weight when erect, reveals a more advanced form of bipedalism than Lucy's. The elbow joint also suggests less and less

BELOW
Bigger and heavier than his predecessors, *A. anamensis* doubtless came into contact with Lucy (*A. afarensis*).

dependency on brachiation. He was a taller and heavier creature – about 1.40 m (over 4 ft 6 in) and 40 kg (88 lb) – travelling upright over the humid, arboreal savannah.

Like Lucy, *A. anamensis* was omnivorous, that is, he ate both flesh and plants. His feeding habits became truly opportunistic as he took whatever his environment offered. Clearly, the freshest prey was the best, as his digestive enzymes were poorly equipped to deal with meat that was at all stale; however, this did not prevent him, when necessary, from battling over a meal with scavengers like hyenas and wild dogs – or even with big cats such as the leopard or the sabre-toothed tiger. He was robustly built, and intimidated other predators sharing his territory.

Australopithecus anamensis began the long walk down the road towards humanity, imposing his presence little by little as a competitor. His feeding habits, perhaps, and his growing ability to walk erect were his chief assets in exploring his environment. He travelled longer distances between rests, and his area of influence extended to eastern Africa. Some experts believe he could even have been our real ancestor. But, for the moment, this is contentious, as the remains unearthed are insufficient to permit a definitive conclusion. All the same, *A. anamensis* looks a very likely candidate.

What teeth can tell us

Teeth can give us very precise clues about the feeding habits of fossilised creatures. Feeding leaves striations and wear marks on the tooth surface, from which experts can deduce the creature's diet. Two methods of analysis are used in tandem, one morphological, the other biochemical. External morphological observations deal with shape, distribution of enamel depth on the crown, and surface wear marks. Internal examination is conducted with scanners and X-rays. Lucy's teeth, for instance, are typical of a predominantly vegetarian subject, with thick enamel, and striations left by the mastication of tough plant matter or fruits. On the other hand, the teeth of *A. anamensis* exhibit different marks: his jawbone operated in a more vertical plane than that of other australopithecines.

Biochemical analysis has provided palaeoanthropologists with new tools, allowing them to isolate 'markers' left by various foods in both bones and teeth. For example, the strontium/calcium (Sr/Ca) ratio decreases as the proportion of meat in the diet increases. Similarly, the relationship between the two carbon isotopes 12C and 13C enables us to determine the type of vegetable matter consumed by the creature during its lifetime. Finally, further information can be assembled by studying animal remains littering the creature's habitat: bones crushed to extract the marrow or showing traces of dismemberment are proof we are looking at a carnivore. Unfortunately, plant debris – seeds, roots, leaves – degrades much more quickly, and so rarely comes down to us.

With teeth more like those of humans, a more
varied diet and a greater sense of curiosity,
A. anamensis could just be the ancestor of modern Man.

THE AFRICAN 'BUSH'

In the arid areas of central, eastern and southern Africa, *Australopithecus* developed along different lines, making the experts' task more difficult, if more intriguing.

Between 1924 and 2000, palaeoanthropologists discovered what amounted to a whole family of pre-humans. They refer to this as a 'bush': a flowering of several more or less parallel branches dating to successive periods between 4.5 and 2.5 million years ago. So far, it is impossible to state with certainty what was the common ancestor of these creatures or which has a claim to be the founder of the human race. They all combine archaic and human-like characteristics, proof that the evolution of these species was not linear. Some scientists even believe that certain features, like bipedalism, went into regression during this vital stage in our history.

'Bush' is a very apt term. Besides *A. anamensis* and *A. afarensis*, present knowledge indicates that branches of the genus *Australopithecus* included *A. africanus*, *A. bahrelghazali* and *A. garhi*. *A. africanus* was the descendant of the Taung child, discovered by Raymond Dart. Numerous individuals, either contemporary with or pre-dating him, were then brought to light at Sterkfontein, Kromdraai, Makapansgat and Swartkrans in South Africa. At the time of writing, a very complete skeleton is being excavated from a South African cave at Sterkfontein, where it is entombed, as if in a natural shell, in a rift in the rock. But the remains already retrieved display fascinating characteristics, like bipedalism. Elsewhere, in a cave at Swartkrans, another *Australopithecus* has been discovered with jaw marks on the skull, suggesting the body was carried off by a leopard either as a kill or carrion.

Louis and Mary Leakey, British researchers based in Kenya, together with Camille Arambourg and Yves Coppens in Ethiopia, have all been engaged in this massive 'bone-rush'. During the 1950s and 1960s, but this time along the East African Rift, they unearthed other remains of australopithecines; some of these specimens were originally baptised *Zinjanthropus boisei* and *Z. aethiopicus* before being absorbed into the genus *Australopithecus*. But most noteworthy was Mary Leakey's discovery (at Laetoli, in northern Tanzania) of three impressive series of footprints extending over a few dozen metres and 3.6 million years old. This 'trackway', evidence of two adults and a youngster walking erect, is one of the most poignant vestiges of the era.

In 1995, when he picked up the anterior section of a jawbone in the Chadian desert more than 2,500 km (1,550 miles) from the great divide formed by the Rift, Michel Brunet, the discoverer of Toumai, suddenly became aware of the immense territory over which the australopithecines had wandered. The owner of this jawbone is technically referred to as *Australopithecus bahrelghazali* – '*Australopithecus* of the River of Gazelles' – but the world knows him under the nickname of Abel, a homage to a deceased colleague of Brunet. Despite his small size, his teeth demonstrate that he was an australopithecine, living more than 3 million years ago.

Finally, in April 1999, the middle valley of the Awash, which had already produced Lucy, provided the team of Berhane Asfaw with a new australopithecine; this was *A. garhi*, 2.7 million years old. The main interest of this specimen is that it belongs to the period when *Australopithecus* was beginning to disappear and the first true Men were emerging.

In any case, this explosion of diverse species, classified by scientists as either 'robust' (*A. aethiopicus*, *A. boisei*, *A. robustus*) or 'gracile' (slender-bodied, like *A. afarensis*) points to the enormous difficulties involved in unravelling all their interrelationships, direct or indirect, not to mention the precise dates of their various physical modifications.

ABOVE
Their already precarious existence doubtless discouraged *Australopithecus* males from excessive fighting. In most cases, an intimidating stare would suffice to deter a less determined rival.

Clues from bones and teeth

Remains of fossilised bones and teeth do not simply provide us with anatomical clues. We can also use them to gain a reasonable picture of their owners' social lives. One of the main findings is that there was a distinct sexual dimorphism, that is, difference in physical attributes between males and females. Sometimes, however, the evidence can be misleading, the truth being that we are dealing with two different species. The males were taller and possessed more powerful canines, indicating that they were the dominant figures. Their superior physical characteristics enabled them not only to hunt and attract mates, but to challenge males from outside the group. The conclusion is therefore that theirs was a society based on the 'dominant–dominated' system, as exists today among chimps and bonobos. Groups in which all the males are related develop less competition.

THE LEGACY OF AUSTRALOPITHECUS

The main contributions of the australopithecines to the story of evolution were their experiments with bipedalism and the first use of stone tools. But most of these creatures disappeared leaving no descendants.

Not all the species of *Australopithecus* were our ancestors. What is more, some experts question whether all of them disappeared without trace or if one species became our progenitors as a result of adapting to new changes in the environment. Once again, climate changes left their mark even on the future of these creatures. However this may be, almost all varieties of this species disappeared without descendants. But the phylogenetic relationships between these species and ours are as yet unclear and opinions remain divided.

Experts cannot agree whether there is a direct link between *Australopithecus* and Man. Martin Pickford, from the Collège de France, thinks that Orrorin began the evolution process leading to humanity, while the australopithecines left no progeny. The problem is that, in our present state of knowledge, a large gap exists when it comes to the link between Orrorin and the first representatives of the genus *Homo*, a grey area waiting to be coloured in by the discovery of other remains from this key period.

And yet the legacy of the australopithecines is considerable, notably their considerable progress in bipedalism, though we need to remember that this was the result of continuous experiment. Some authorities do not hesitate to speak of an evolutionary pause or 'stasis' when they compare the walking strategies of these creatures: some – the older groups – display more 'modern' characteristics than others who are, conversely, more recent.

It was *Australopithecus* who first, as far as we know, issued a challenge to Nature by learning how to make stone tools. Nothing very elaborate, to be sure, if we compare them with the efforts of the first species of *Homo*. But the action of striking a stone with another to impart to it a definite and useful shape dates to 3.5 million years ago, with traces of small, knapped quartz tools found some time ago in Ethiopia, in the Omo valley. Admittedly, we do not know which type of hominid leapt this intellectual barrier. On the other hand, *A. garhi* can be associated with flint tools whose cutting edge definitely served to carve up animal carcasses.

The study of bone remains enables us to estimate not only the age at which their owners died, but also their life expectancy: 20 to 30 years, on a level with that of chimpanzees and other great apes. If Lucy died at 20, other australopithecines may have lived longer. Under pressure from environmental factors and internal or external relationships (food or sexual competition, for instance), they organised themselves into tightly knit groups. In Ethiopia, Maurice Taïeb, Donald Johansen and Yves Coppens excavated, from a single site near Hadar, fossils of a dozen individuals, all contemporaries, but of differing ages. The collection from the site was baptised the 'First Family'. We have to suppose that such a group's social response to the environment demanded complex processes of learning and cultural behaviour, all indicative of an intellectually richer organisation.

Finally, the mobility of *Australopithecus* demonstrates the way he had developed as a predator and explorer over more than 2 million years. His immense range, from Chad to southern Africa, suggests a creature already enjoying a certain capacity for reflective thought resulting from an in-depth knowledge of his environment. Hand in hand with this went an ever-developing opportunism, enabling him to harvest all the lavish resources of the world he occupied.

OPPOSITE
Additional proteins derived from meat allow the brain to develop and acquire new capacities.

RIGHT
It was undoubtedly by chance that australopithecines discovered, some 3.5 million years ago, that stones could be chipped and shaped. But some time would pass before hominids acquired this skill.

Where is our ancestor?

Classical theories of evolution have accustomed us to consider at least some species of *Australopithecus* as the ancestors of the genus *Homo*. But the discovery of Orrorin suggests an alternative scenario, already adopted by some experts. According to this, Orrorin was the progenitor of the human race, after passing through a form discovered in Tanzania; the various species of australopithecines formed a parallel branch, which disappeared on the emergence of the first specimen of true Man: *H. habilis*. Even so, several questions remain unanswered. Fresh discoveries may confirm the theory or – who knows – give rise to one or more new ones.

HOMO HABILIS

THE FIRST TOOLMAKER

The first true representative of the human race enjoyed intelligence and conscious thought, gifts that would henceforth guide his every step, inspiring him to fashion elaborate tools, to pass on his knowledge to his children and, in short, to conquer the Earth.

OPPOSITE AND LEFT
His more complex brain, with its increased effectiveness, enabled *H. habilis* to feel more at home in his environment.
He exploited this new capacity by discovering different foods and making his own tools.

BRAIN AND CONSCIOUSNESS

Homo habilis still shared some archaic characteristics with his predecessors. But the true difference lay in his brain.

On the arboreal savannahs of eastern and southern Africa, a new hominid appeared 2.5 million years ago. His earliest remains were discovered by Louis and Mary Leakey at Olduvai in Tanzania in 1960. Louis Leakey, Philip Tobias and John Napier named the new find *Homo habilis* (the 'handyman') for, among the remains (skulls – more or less complete – jawbones, teeth, foot and hand bones) often scattered about a dwelling site, teams also came across stone tools. *Homo habilis* was more skilled in fashioning tools than his contemporaries, the last of the australopithecines, demonstrating an excellent knowledge of his geological environment and a technique that owed nothing to chance. He 'knew' what he was doing.

What really set him apart from his forebears lay hidden inside his skull. His brain was bigger – 600 to 750 cm^3 (approx. 37 to 46 cubic inches) – and, most important, more complex. Examination of the inner surface of the skull (endocranium) reveals development of the frontal and temporal lobes that governed the way he would 'see' the world and his social relationships. These lobes are supplied by a so-called middle meningeal vessel with a frontal network denser than in older hominids, ensuring a greater flow of blood and thus more effective thermoregulation of the brain.

What is more, the right-hand area of his brain includes two zones, Broca's and Wernicke's areas, governing respectively the production and comprehension of articulate sounds. We might interpret this to mean that, in addition to signs and other gestures, *H. habilis* was already capable of meaningful communication with his fellows in the form of an embryonic language. For now, this remains a matter of debate, as chimpanzees also possess such zones. Yet the larynx of chimpanzees and other great apes is sited higher in the throat, with a consequently reduced resonating chamber at the back of the mouth; in any case, the larynx is not an organ that survives in the fossil record. As things stand, then, the only safe conclusion is that these early Men were equipped with the structures necessary for spoken language, and may possibly have been capable of using them.

From a strictly physical point of view, *H. habilis* was of average size for hominids of this period – between 1.10 and 1.40 m (3 ft 7 in and 4ft 7 in) – and weighed from 30 to 40 kg (66 to 88 lb). His face was flatter, while his limbs show he was adapted both to bipedalism and tree-climbing. In Kenya and Malawi, during the same era, there existed another species of *Homo*, more human in his posture and ability to walk erect, with an even larger brain: 700 cm^3 (43 cubic inches). He was *H. rudolfensis* – 'Man from Lake

Fossilised brains?

When palacoanthropologists draw their conclusions about the brains of prehistoric Man, they do not, of course, have access to the brain itself. This organ is composed of soft tissue, which decomposes after death before other body parts. Our knowledge of ancient brains derives exclusively from making casts of the inner surface of the skull or endocranium. What happens is that the bones enclosing the brain preserve on their internal surfaces the imprints of certain veins and arteries controlling its blood supply, as well as those of special zones involved, for instance, in speaking and comprehending language. These traces enable researchers to determine the brain's degree of complexity and thus to form an estimate of its capacity for conscious thought.

Rudolph' – after the place where an incomplete skull was discovered, modern Lake Turkana. The jury is still out among specialists, who are striving to draw up precise definitions of these two species who flourished side by side between 2.5 and 1.7 million years ago. Nonetheless, in *H. habilis* we appear to glimpse the beginning of consciousness in the sense we understand it: awareness of oneself and others, with all its social implications.

OPPOSITE
Doubtless by chance, this hominid has just discovered flint-knapping. The process would be infinitely repeated; similarly worked stones are found in every region inhabited by early Man and the last australopithecines.

THE TOOLMAKER

Using tools was not new: chimpanzees do it routinely. The real miracle was that Man began to make his own.

Pre-humans and chimpanzees shared the same natural abilities for millions of years, taking objects occurring naturally in their environment, like branches or tough plant fibres, and using them as tools. *Homo habilis* and some australopithecines took the innovative step of deliberately transforming an object – usually a stone but sometimes pieces of wood – to manufacture a functional object, that is, one with a specific purpose. This suggests observation, learning and manual dexterity: the toolmakers had mastered a series of techniques and planned a use for what they made. In short, they could think, and had developed an authentic culture.

The first true stone tools are more than 3 million years old (Omo valley). In the gorges of the Olduvai, in Tanzania, or at the Lokalei site in Kenya, thousands of these items came to light between the early 1960s and the end of the twentieth century. Clearly, even in their own distant epoch, the first representatives of humanity had mastered extremely efficient techniques using 'pebble tools' to shape stones and turn flakes of flint, basalt or obsidian into blades and other implements.

Nobody yet knows 'who made what', as Yves Coppens was already pondering back in 1982. But evidently these tools, developed with growing precision over 700,000 years, were designed for distinct purposes. They have been discovered in the immediate proximity of piles of animal bones. These bones have been broken up to extract the marrow, while some have obviously been separated from one another using 'choppers', large stones specially prepared with finer, sharper edges to dismember cadavers before carving up the flesh. Also among the finds are stone planes, chopping blocks and percussion tools.

The uses to which tools were put can be discovered from the techniques of traceology. Using electron microscopes, experts are able to show that, like teeth, stone tools preserve marks of wear and abrasion on their working surfaces resulting, say, from cutting up an animal or slicing through tough plant stems. Again, reconstructing a nucleus by means of the flakes separated from it by the prehistoric artisan allows us to elucidate the technique in question. *Homo habilis* was fully conversant with the nature of the materials he worked with, and was able to detach flakes of a predetermined form according to the texture of the stone and the angle and direction of the cut. Toolmakers' 'workshops' discovered near large deposits of flints also indicate that the first men did not work at random: they knew where to find the best stone, meaning that they had a rational understanding of their environment.

BELOW
'If stone chipped by stone can cut my hand, it must be able to cut other things as well!' This must have been the deduction; it shows how far the first men had begun to ponder the meaning of things around them.

OPPOSITE
First experiments to reproduce the action that turns stones into useful tools.

DOUBLE PAGE OVERLEAF
Armed with his discoveries and his larger brain, *H. habilis* is emboldened to compete for food with the big predators of the savannah.

Palaeontologists speak in this connection of the 'Lower Palaeolithic'. Situated at the start of this era, the Oldowan culture – named after the Tanzanian site of Olduvai – can be dated to 1.8 million years ago. The most famous site is Melka Kunture, in Ethiopia. It was indeed a culture worthy of the word, which would develop and accompany the first humans as they set out from Africa upon their centuries-long conquest of the Earth.

Tools by the metre

Beginning with pebbles sharpened to make fine arrowheads, men ceaselessly, over a million years and more, refined their techniques and developed ever more effective tools. An excellent way to illustrate their technical progress is to measure the length of blade obtained from 1 kg (2.2 lb) of stone at different periods in prehistory. The results are amazing.

From each kilogram of 2-million-year-old stone, 10 cm (4 in) of blade was produced. This rises to 40 cm (16 in) at 500,000 years, 200 cm (79 in) at 50,000 years and 2,000 cm (790 in) at 20,000 years ago! This simple comparison demonstrates not only how techniques improved, but that toolmakers made better and better use of their raw materials.

The reconstruction of cores reveals that the actions of striking stone upon stone were carried out by right-handed artisans since the earliest times. Finally, from the dimensions of stone tools we can see that they were designed for men's or women's hands about the same size as our own.

Day after day, *H. habilis* tried to puzzle out the mysteries of the world about him. Slowly but surely his consciousness was awakening.

OPPOSITE
Doubtless to recapture the security he had enjoyed in the trees, *H. habilis* tried to construct shelters made of branches.

Other hominids, like this *H. ergaster*, would copy and improve the design.

THE FIRST DWELLINGS

As they became more adept at walking upright, *H. habilis* and *H. rudolfensis* had to leave the safety of the trees and find refuge at ground level. For the first time, our ancestors built themselves shelters.

The earliest men were nomads. They wandered over the savannah in groups of often several dozen individuals, pooling their resources to exploit every possible source of food. As they found it easier to walk upright and the arid zones once more began to advance, they lost the possibility of taking refuge in the trees when danger threatened or if they simply wanted to sleep. Unlike other animals, our forebears were not content merely to sleep on the ground; they therefore needed to innovate.

Their ability to think, already harnessed to toolmaking, would also enable them to escape this somewhat perilous situation. Initially, they made use of environmental features like cliffs or ledges under rocks to keep safe at night. By day, however, they let their imaginations run free.

Various stone circles, found especially in Kenya and Ethiopia, suggest that these hominids used simple geometrical shapes to support vault-like assemblages of branches – to all intents and purposes small huts. It is as if they wished to re-create the security of the tree canopy by transferring it to ground level. But, more importantly, these sleeping areas, sometimes raised and levelled, as at Gombore in Ethiopia, seem to indicate a desire to clear a well-defined zone for more or less permanent occupation.

Did these early dwellings, which date to between 1.8 and 1.3 million years ago, offer reasonable protection against carnivores? Did hominids employ defensive strategies like placing thorny branches around their shelters? The answer is not easy to determine, as, of course, the vegetal matter that formed the raw material for these dwellings has vanished. But ethnological comparisons with modern African peoples suggest this may well have been the case.

On the other hand, these habitats have left the indelible trace of a new social organisation among the hominids. Such constructions presuppose a communal effort by the group, employing a distinct division of labour, to ensure its safety and its food supplies. That early Man engaged in specialisation is evident from the traces of areas set aside for stone-chipping or butchery. Again, the fact that these are sited near lakes or rivers reveals that the itineraries of *H. habilis* and *H. rudolfensis* were closely associated with water sources. Later dwelling sites, on the other hand, were perched on river terraces, suggesting that our ancestors had devised means of transporting water, but evidence of this has also been erased with time.

The tendency to build sites near water involved several dangers for prehistoric populations, such as floods, mud slips, or herd movements of large animals. But for us it has the immense advantage that it helped preserve some traces of their primitive dwellings.

OPPOSITE AND RIGHT
The fact that *H. habilis* engaged in more or less successful experiments – in other words learning processes – is an important reason to regard him as the first true Man.

THE COMMUNICATION OF KNOWLEDGE

Making stone tools, building shelters from branches and using specialised labour supposes memory, learning and the exchange of information. In short, a transmission of knowledge from generation to generation. Teaching thus became a pivotal factor of culture.

The world's first school consisted, perhaps, of a flint-knapper teaching one or more youngsters from his group how to make tools. *H. habilis* and *H. ergaster* owed their success partly to this ability to communicate information and learning (i.e. knowledge) from one generation to the next. Their more highly developed brain, with its denser network of blood vessels, empowered them to carry out this transmission on which the permanence of the group depended. Gestures, mimicry, example and, no doubt, an elementary form of language made it possible to hand down skills beyond a single lifetime.

It is axiomatic in modern societies that nothing replaces experience for carrying out tasks involving precision. But this is meaningless without the possibility of passing on the knowledge to one's group or humanity as a whole. The Shungurian and Oldowan cultures were, in this respect, authentic Palaeolithic civilisations which began to multiply across the entirety of Africa 2.5 million years ago. The Acheulean culture relayed that of Olduvai, with its artisans fashioning an even more sophisticated tool – the biface. This period occupies a long time in prehistory: from 1.6 million to 300,000 years ago.

Apprenticeship was also a means of instructing the younger generation in state-of-the-art techniques – stone-chipping, shelter-making, hunting and fishing – at any particular moment. Here was the 'core curriculum' from which some individuals learned advanced techniques, honing their skill still further, thus reinforcing their importance within the group and, above all, ensuring its overall survival. The brains of the earliest humans were therefore continually occupied with new challenges. New neuronal connections formed, imagination and reflection

There is continual progress in division of labour, sharing of food and tools and organisation of the clan. Gradually human society emerges, with all its rivalries and conflicts.

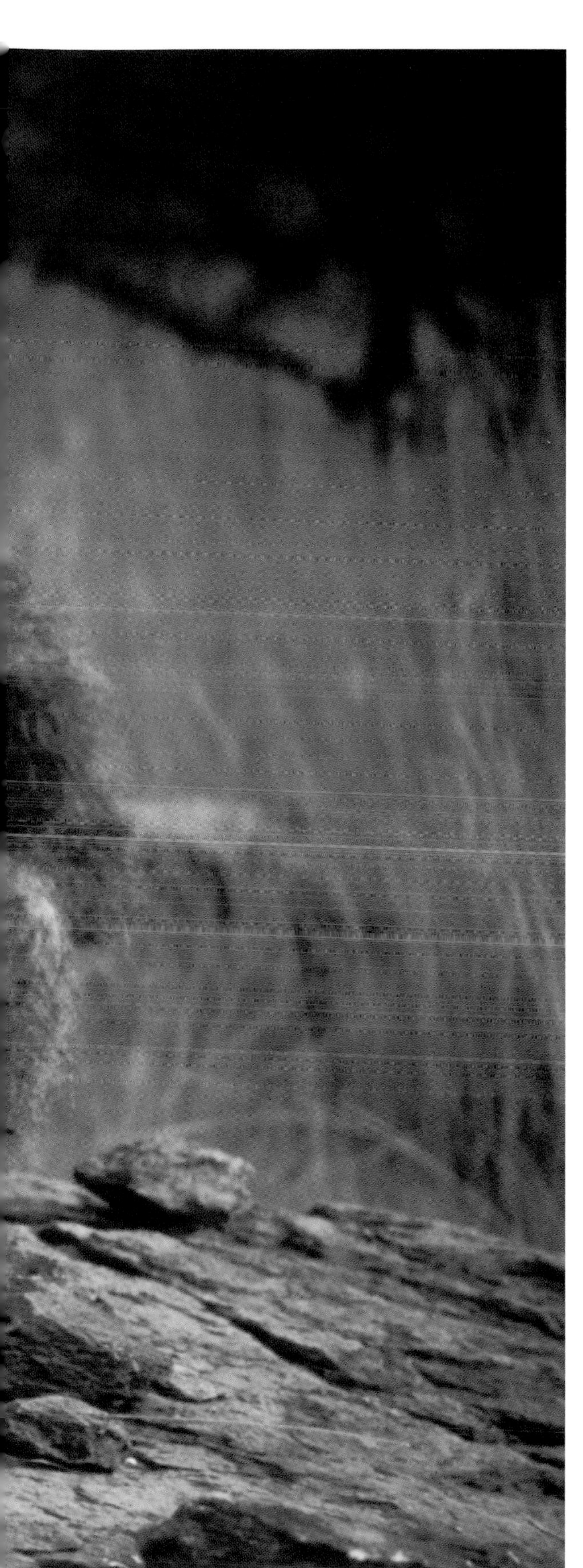

LEFT
The first experiments of *H. habilis* in using language would prove of enormous benefit to the development of his culture. This was the beginning of teaching and the communication of knowledge.

reinforced each other, leading to the production of tools that were of superior quality or easier to handle.

All the same, the basis of such progress remained unchanged. The fundamental appearance of the chopper or the biface had not altered in a million years, even if their finish and precision had improved. The same is true of the hand axe, the iconic creation of the African Acheulean culture, whose technical design hardly varied when it was transmitted from eastern to central Africa, to Maghreb and the Iberian Peninsula.

The range and rapidity involved in the spread of these techniques also suggest that there were contacts between different groups of hominids, allowing them to acquire new toolmaking skills. On the other hand, cohabitation between the Oldowan and Acheulean civilisations seems to demonstrate that certain groups preserved more or less archaic techniques, while others wholeheartedly adopted the new methods before improving on them and exporting them in their turn. From China, where the Renzidong site has yielded stone tools dating back 2.1 million years, to southern Europe, this technology was disseminated from Africa, accompanying the astonishing expansion of the hominids across every continent.

HOMO ERGASTER

THE HUNTER-GATHERER

Standing tall, *H. ergaster* dominated the savannah, and his superior brain allowed him to make a wider range of tools. He was able to leave Africa and set off to conquer the planet – if this conquest had not already been begun by his predecessors.

THE BIRTH OF SOCIETY

As his brain continued to grow, *H. ergaster* not only produced incredible inventions but discovered new feelings and new social relationships.

With a brain capacity of 900 cm^3 (55 cubic inches) – almost as much as that of some modern *H. sapiens* –, *H. ergaster* enjoyed unparalleled thinking power. Fresh neural networks arose, developing specific zones of the brain, such as those associated with language, facilitating and extending an individual's interactions with the group. There was an enlargement of the frontal lobe, seat of the emotions, and the parietal lobe controlling the senses. Most importantly, the functions of these areas became more and more efficient.

Physically and intellectually, *H. ergaster* represented a real evolutionary leap forward in the human line. But the earliest of his species lived contemporaneously with the last

OPPOSITE
'He is crying. What makes water run down his face?' With social bonds strengthening and becoming more complex, *H. ergaster* begins to be aware of the function of time and the connection between events. Perhaps, for the first time, he ponders the meaning of life.

BELOW
The clan's camp is now well organised. The notion of a hierarchy linking individuals in a group leads to structuring of space.

australopithecines and scattered groups of *H. habilis*. Palaeoanthropologists are thus faced with a difficulty in determining the exact links between the different branches. For instance, *H. ergaster*'s height, as measured from skeletons, is very close to that of modern Man, which is not the case with contemporary species.

What is more, the size, and particularly the functioning, of his brain imply considerable expenditure of energy in comparison with the rest of his body. The brain alone absorbed more than 25% of the body's total energy while amounting to only 2% of an individual's weight. This would explain the search for a more calorie-rich diet available in all seasons and all places: fresh meat.

His large brain and the thinking-power it generated also triggered revolutionary social relationships: *H. ergaster* was aware that he formed part of a community, and that the survival of this community hinged upon the behaviour of individuals. This new group ethic led him to become dependent upon others and in turn to offer them help. For the first time the animal kingdom was confronted with notions of solidarity and belonging; however poorly these were articulated, they were seminal concepts in the development of humanity. The organisation of the group was also radically modified; by around 1.8 million years ago territory was probably structured around a base camp. Such an organisation also presupposes a precise social architecture, including, doubtless, a division of labour according to gender, with the men doing the hunting while the women collected small game and plant foods.

The developing awareness of *H. ergaster* led him also to ponder death and the precarious nature of life. The very existence of nuclear families and organised groups implies that of feeling, of emotions common to members of the same family unit, the same clan. The sum total of these mental processes forms one of the pillars of the hierarchical and structured society that sets the human species apart.

Clan structure and population distribution

Today there are over 6 billion of us on the planet, but it is likely that 2 million years ago, *H. ergaster* numbered only some tens of thousands. By studying present-day societies, such as the Inuit hunters of Greenland, we can estimate that early *H. ergaster*, the hunter, lived in groups of 20–30 individuals. For reasons of survival, whenever population levels rose unacceptably, a small group would detach itself and take up residence a short distance away.
Where human fossils are sufficiently numerous in the same deposits of the same era to be statistically significant, the population can be estimated at around one human per 10 km^2 (about 4 square miles) – approximating to the density of the aboriginal population in certain areas of Australia.
It appears, then, that Man began his colonisation of the planet in small, organised groups.

HOMO ERECTUS

MAN ON THE MOVE

Man no longer merely walked upright, but had learned to think. His social organisation and technology held before his eyes the keys to the Earth.

OPPOSITE
Larger and boasting a bigger brain, *H. erectus* was becoming more human with social behaviour and faculties much more like our own.

BELOW
Armed with his greater intelligence and new weapons, *H. erectus* – 'Upright Man' – became a formidable hunter.

THE SKULL: ARMOUR FOR THE BRAIN

With a brain now twice the size of his cousins, the great apes, *Homo erectus* advanced with giant steps down the road to modern Man.

H. erectus – Upright Man. This new hominid, who also appears to have emerged in Africa, was tall: between 1.60 and 1.80 m (5 ft 3 in and 5 ft 9 in), almost the same height as us. But it was his brain which set him apart from *H. habilis*. Its capacity of some 900 to 1,000 cm^3 (55 to 61 cubic inches) was double that of his cousins, the great apes. Once again, this development was part of a cycle; he adapted to an environment and then began to evolve anew. He had no choice, therefore, but to alter his diet and thus become truly omnivorous.

The mastery of fire some 500,000 years ago, bringing with it the invention of cooking, must have contributed in some degree to Man's evolution. The proteins and fats he absorbed as a result allowed him to store increased quantities of energy. This would be used by the brain, which in turn grew even bigger.

The skull of *H. erectus* had very thick walls: up to 1.3 cm (0.5 in) in places. His forehead was receding, the face prognathous (i.e. sloping forward, with no chin, and, above the eyes, a bony ridge). The reinforcement of the temporal and occipital bones surrounding the brain helped protect this vital organ.

H. erectus had larger, more powerful teeth than our own, indicating that his diet consisted principally of raw meat and plant matter. The bones of his arms were very robust, but his posture and gait were still quite recognisably modern.

These evolutionary advances empowered Man with new intellectual possibilities. His hunting became more efficient, his speech more articulate; he mastered the art of making fire and communicated his knowledge to wider and wider circles. All in all, the development of his brain enhanced the sophistication of every aspect of his behaviour.

Yet the brain possesses one important defect: it is fragile. Its vertical alignment leaves it at the mercy of temperature variations – as we know well enough today – that can cause illness like sunstroke and even death. The hair offers, of course, a reasonable protection, but not always enough. In fact, the solution arose from a modification of the blood vessels in the brain, allowing more heat to be transferred away from the head.

BELOW AND OPPOSITE
H. erectus killed to eat. He had learned to procure the only food available throughout the year: fresh meat.

The prime weapon of *H. erectus*, the predator, was his organisation.
To compete with lions and wolves, he exploited the strength of the group, meticulously preparing his forays in advance.

OPPOSITE
Language developed only gradually. With improved communication between group members, exchange became easier, particularly that of knowledge.

TOWARDS AN ARTICULATE LANGUAGE

Raw sounds became words, paving the way for more sophisticated communication. The larynx becomes the key to the making of Man.

Continuous adaptation to arid conditions since the events of the Rift Valley led to numerous anatomical modifications. One of the less visible, but most important for our later history, lay in a modification of the respiratory system which resulted in the larynx occupying a lower position. Man is the only vertebrate to possess such a deep-set larynx. This unique factor, in combination with the development of the vocal cords, produced a sort of sounding box between the latter and the mouth.

The tongue also acquired greater mobility owing to the extra space resulting from deeper placement and reduction in size of the mandibular bone, sited behind the incisors. In addition, the new method of body-cooling through sweating meant that men were no longer forced to breathe by panting. Now they could control their respiration without regard to the external temperature. This is one of Man's most vital assets, forming the basis of his ability to utter articulate language – impossible without total control of the breath.

A sounding box, a more mobile tongue and control of his breathing-rate allowed Early Man to speak, and his language soon began to resemble ours with ever more elaborate utterances. Study of his endocranium reveals the presence of Broca's area, the zone in the frontal region controlling speech production, already present in *Australopithecus* and the great apes, but now much enlarged and associated with the similar Wernicke's area in the occipital region, essential for the understanding of language.

Examination of casts made from prehistoric skulls (*see p. 55*) shows that the left side of the brain was more developed than the right, a characteristic also found in apes. From this asymmetry we learn that the temporal planum, the zone of the temporal lobe active during linguistic processing (expression, comprehension) had already acquired a much greater complexity.

Of course, the language of Early Man was nothing like as sophisticated as ours, and *H. erectus* most likely encountered difficulties in enunciating certain sounds. But he would rapidly improve until he spoke with the complexity we are familiar with today.

The reason why this increasingly more elaborate means of expression was of such strategic importance to future branches of humans was that it enabled them to exchange ideas more easily, with a consequent mutual gain in knowledge and experience. Communication between humans is fundamentally different from that between other animals. It is not based solely on demands, but also upon a desire to share one's representation of the world: a level of cognition which made us what we are today.

Grammar maketh Man?

The areas of Broca and Wernicke are vital for the production and comprehension of language. Scientists have demonstrated that even chimps possess them. What ultimately distinguishes human language from that of animals is its complexity and its capacity to form sentences. It is true that some birds, for instance, can utter hundreds of different sounds, each corresponding to a precise situation. But only Man, with his articulate speech-patterns based on complex vocabulary, grammar and syntax, can link sounds together to form words and eventually sentences. According to the way these sounds are associated, their meanings can vary.

BELOW AND OPPOSITE Considered during the Middle Ages as magical ('thunderstones'), the tools produced by *H. erectus* became ever more varied. Bifaces (handaxes) now joined various scraping and striking implements.

EARLY MAN AND HIS TOOLS

H. erectus had enormously improved his methods of making stone tools. He now possessed a whole range of weapons and implements for every necessity, but for acquiring food supplies in particular.

The earliest bifaces, technically known as proto-bifaces, appeared with *H. ergaster*, some 2 million years ago (*see p. 74*), but the biface in its strictest sense was the invention of *H. erectus*.

The biface (i.e. cleaver or handaxe) had a double edge, as its name suggests, and was one of the most remarkable items manufactured by prehistoric Man. Typically it was about the size of a human hand, and oval or triangular in shape. The two faces formed convex surfaces whose intersection resulted in what was for most of its length a straight working edge. A biface could be fashioned from nodules or flakes of flint, quartzite, polished sandstone or other suitable material.

The dimensions of bifaces differ considerably; the largest exceed 20 cm (8 in) or so, the smallest measure only a few centimetres. Shape and thickness are also variable. This extreme diversity reveals not only a definite mastery of the production of these tools, but also that *H. erectus* had designed a whole range of tools for his needs.

The biface may be considered as the 'Swiss army knife' of prehistory. It was an all-purpose tool and occurs at every period, from 1.5 million years ago to much more recent times, around 500,000 years ago. Rough and ready at first, it took on shapes described as 'laurel-leaf' or 'willow-leaf' when *H. sapiens* perfected the art of stone-working. The biface had by then become extremely fine, a few millimetres thick, and must have performed an artistic or symbolic as much as utilitarian function, given its apparent fragility. Despite the many traces of toolmaking found from northern France to eastern Africa, mystery shrouds the origin of the biface and how it evolved from the pebble-tool. But we are forced to admit that such tools demanded an understanding of symmetry, one of the hominids' most elegant conceptual discoveries.

It was at this precise moment that men stopped copying Nature – pebble-tools were reproductions of stones degraded by natural forces – and impressed their own stamp upon their materials. Their experiments were no longer limited to questions of efficiency or usefulness: they had ventured into the hitherto unexplored regions of harmony and beauty.

The experience of Early Man had become irreplaceable. He now had a choice of materials, and techniques for extracting and working them.

OPPOSITE AND RIGHT Relationships within the clan grew more complicated, the tasks to be done more numerous. Efficiency demanded the division of labour.

THE DIVISION OF LABOUR

Within the *H. erectus* clan, individual members began to specialise. Precise relationships developed, while a sense of kinship grew ever stronger.

Careful study of clues left at sites occupied by *H. erectus* reveals how areas were set aside for specific purposes. Near the sleeping quarters would be the place where the clan ate, and what amounted to a workshop for making lithic tools. These clues, often tenuous, indicate a relatively elaborate organisation in the way tasks were distributed. In the case of slightly later *H. erectus*, some 500,000 years ago, we even find vestiges of a hearth where a fire was maintained over varying lengths of time.

Whatever their environment, the men hunted to bring back meat for the group and the women played their part by gathering fruit and berries or using digging-sticks to unearth a variety of roots. Group members met up again for meals and rest; meanwhile one of the men would be busily engaged in making the tools essential for community tasks.

With the growing sophistication of tools, techniques became more complicated and time-consuming. The community needed every available hand – and skilled hands at that. Choices had to be made about extraction, knapping, shaping ... The experience of the older members was now priceless.

The communication of knowledge had become vital for the continuity of the species and accordingly received far more attention than in the days of *H. habilis*. With the new means of expression made available by the acquisition of articulate language, a kind of open-air school would grow up as the veterans taught the youngsters to recognise suitable materials for toolmaking. The new generation learned to work weapons and tools (sometimes, as today, improving on their elders' techniques), imbibed valuable lessons about hunting, or helped the adults prepare the meals.

Unlike his modern counterpart, the young *H. erectus* never experienced the crisis of adolescence. Teeth studies of these hominids reveal that the period of growth roughly approximated to that of chimps or australopithecines: very rapid and swiftly completed. During the later course of evolution a spectacular physical modification occurred, leading, at around 800,000 years ago, to longer childhood; the benefits were that youngsters had more time in which to learn survival skills and make full use of their brain capacity.

As for relations between groups, it is hard to tell whether conflict or co-operation was the rule. But examination of human remains from the Palaeolithic period show that combats were not rare. The myth of peaceful primitive communities appears to be one of those utopian visions of the brotherhood of Man promoted by eighteenth-century philosophers.

BELOW AND OPPOSITE
The women probably did not often take part in big game hunts, but they played a vital role in the community, gathering plant foods and guarding and teaching the children.

DOUBLE PAGE OVERLEAF
The hunt over, a man relaxes in the security of his clan.

The model of 'relic' societies

The vast majority of deductions about our ancestors' behaviour derive from observations of present-day populations. This is a principle known as 'actualism'. There still exist certain societies (erroneously) labelled 'relic' or 'primitive' populations because they practise archaic or primitive techniques. Living in a virtually self-sufficient manner by hunting, fishing and gathering vegetal matter, in harmony with Nature – taking only what they need – such populations can be considered as balanced. They can be viewed as models of tradition, culture, male/female relationships or the division of labour. This leads to the conclusion that the first men to form organised societies, like *H. ergaster* and *H. erectus*, lived in a fashion analogous to the Inuits of Greenland, the Australian Aborigines or certain African tribes still practising a nomadic lifestyle. This handful of societies is often used as a benchmark for gauging ancestral lifestyles.

TOWARDS THE EAST

H. erectus colonised all Asia, except the mountains and regions of inhospitable cold, giving birth to new civilisations.

After *H. habilis* left his African cradle, he settled in Asia and Europe, where he eventually became *H. ergaster* and then *H. erectus*. At least, this is the opinion of some experts; when we come to try and identify the course of human migrations from Africa to Eurasia, to say that things are complex is a gross understatement.

Discovered and described for the first time by Eugène Dubois in nineteenth-century Java, *H. erectus* was initially known as *Pithecanthropus erectus* ('Upright Apeman'), and only later received his present title. Java Man's physical characteristics, indeed, appear to be those of *H. erectus*, especially the skull and the limbs. If the theory of the 'great migration'

ABOVE
H. erectus originated mainly in Asia, but rapidly colonised the whole planet, except, perhaps, for the Americas.

The migrations: two opposing models

Two theories regarding the prehistoric migrations of *H. erectus* are currently 'head-to-head'. The first, known as the 'regional continuity' or 'multiregional' model, admits the existence of a common stem represented successively by *H. habilis*, *H. ergaster*, *H. erectus* – who diversified throughout most of the planet – and finally *H. sapiens*. But such an interpretation appears simplistic, possibly blinding us to more complicated movements between Africa, Asia and Europe. Currently the majority of palaeoanthropologists embrace the second theory, known as 'single origin'. According to this, the Asiatic branch descended from the last populations of *H. ergaster* who migrated before 1.5 million years ago, taking with them Oldowan tools and techniques which persisted until much more recently. The 'robust' species of Asia were thus but one in reality: *H. erectus*. The representatives of *H. ergaster* left behind in Africa – also known as *H. heidelbergensis* – underwent a rapid expansion throughout that continent between 1.5 and 1 million years ago, then in the Middle East and Europe around 500,000 years ago.

('Out of Africa') is correct, it appears that *H. erectus* colonised Asia very rapidly after his departure from Africa. Worked stones have been found, notably in China and Indonesia, all aged around 2 million years.

China is, in this respect, a vast reservoir of discoveries, as the country enjoyed a favourable climate at the period in question. At Longgupo, in the east of Sichuan Province, a cave-fall yielded two apparently worked stones, a few teeth and a fragment of jawbone, all dating to 1.9 million years ago. It was at Zhoukoudian, 48 km (30 miles) south-west of Beijing that *Sinanthropus pekinensis* – Peking Man – was discovered. He was a variety of *H. erectus* working lithic tools and began inhabiting this area 500,000 years ago.

The five skullcaps and other remains of jawbones recovered at Zhoukoudian in 1927 suffered a bizarre fate. When the Japanese invaded in 1941, the precious fossils were entrusted to evacuating US Marines for shipment to the States. But the marines were taken prisoner and what happened to the crates remains a mystery.

It was doubtless at the time of *Sinanthropus* that a division of human types arose. Arriving in new climates and new environments, Man adapted rapidly in a few hundred years – or a few thousand at the most. Specific physical variations developed, such as skin colour or the shape of the nose or the thoracic cage. So-called 'regional' populations established themselves in every corner of the globe at a period when *H. neanderthalensis* and *H. sapiens* were rubbing shoulders in the Near East and Western Europe.

MASTERING FIRE

The ability to harness fire was nothing short of a cultural revolution; with it, humanity took a giant step forward.

The first traces of fire were found in Swartkrans, in South Africa, and date to around 1.5 million years ago: three pieces of charred bone appearing to indicate controlled burning. Even if this incident can be attributed to *H. ergaster*, use of regular hearths only began to spread with *H. erectus*. What is more, the phenomenon arose virtually simultaneously at both extremities of Eurasia, in China (Zhoukoudian) and in Brittany (Menez-Dregan).

The existence of these hearths is revealed by ash residue and the stones trapped within it; the latter have been cracked and their colour modified by heat. Besides these scorched stones, investigations have also turned up a large number of calcined bones and small quantities of wood charcoal.

It could have been no easy task for Early Man to learn the art of fire-making; he must first have had to overcome the fear that fire inspired in him as it did in animals. Initially, *H. erectus* would no doubt have had to bring back burning branches from forest fires to his shelters or caves. He would feed this fire and make it last as long as possible to keep himself warm and ward off predators. Soon he would learn to cook, after stumbling on the fact that, not only did meat taste better cooked, but, being more digestible in this state, had fewer harmful effects on his stomach.

The first incontrovertible signs of regular hearths date back half a million years. In the Caune (Cave) de l'Arago, at Tautavel in the eastern Pyrenees, no trace of fire has been identified, but, as in modern times with nuclear fission, not everyone acquired the technology simultaneously. At this time, however, many other sites boasted regular hearths: Terra Amata, near Nice; Vertesszölös (Hungary); Orgnac (in the Ardèche), etc.

With the harnessing of fire, Man laid his hands on a formidable tool that freed him from certain natural constraints, such as the absence of light at night, bitter cold in the winter and the

PRECEDING DOUBLE PAGE
Fire brought cheerfulness, warmth and protection. It soon became an indispensable ally of the tribe.

BELOW AND OPPOSITE
H. erectus revolutionised his life by taming fire, which he had once considered as an enemy to be feared. Finally he learnt to make it himself, transforming his whole life.

threat from large carnivores. Above all, fire allowed him to extend his territory. Despite the glaciations and changes in climate, *H. erectus* and his fellow human species now enjoyed a new form of communal or tribal organisation based round the hearth that provided them with light and heat and had thus become an essential catalyst of social intercourse. Culture, in the wide sense of the term, is in part the outcome of Man's taming of fire.

With the acquisition of this new power, Early Man finally bade farewell to the animal world, and fire would accompany him, in many guises, all along the tortuous road of his future development. His whole destiny would be changed by this social revolution.

Fire-making techniques

There are, or were, various ways of making fire. Today we only need to press a button, but it took tens of thousands of years after Man first harnessed fire before he could actually create it himself. One of the first techniques *H. erectus* may have used, and which is still employed by some populations, consists of twirling a stick between the hands on another piece of wood in order to heat it. At the right moment, the fire-maker adds a handful of dry grass or leaves to get the flame going.

It was almost certainly only later that our ancestors learned to produce sparks by tapping a piece of flint on a lump of pyrites and using the sparks to light a small heap of dried grasses or moss. This method appeared around 30,000 years ago, and proved much quicker and more practical, provided that suitable stones were to hand.

YEARS

-200,000 -40,000

NEANDERTHAL MAN

THE OTHER FACE OF HUMANITY

Some European populations of *Homo erectus* were trapped by glaciations. The way they adapted to the cold climate modified their development. Neanderthal Man evolved in a world of his own, establishing a unique form of civilisation.

EUROPE: THE ICE TRAP

In Europe, a cold climate persisted as one ice age followed another. *H. erectus* was trapped by glaciers that isolated the whole continent. As a result he developed along different lines.

The human species is, like all forms of animal life, shaped by climate and its changes. Once again, this variability was at the origin of an important episode in the history of the genus *Homo*. Between 900,000 and 15,000 years ago, there was a sequence of no less than nine ice ages, providing the clearest possible evidence of the devastating effects these fluctuations can have upon our planet. This era witnessed periods of cold, with an average temperature lower than today's, alternating with all-too-brief temperate interglacial interludes.

The first men arrived in Europe more than a million years ago – probably more like 2 million. There they became trapped by these successive glaciations which, in the north-west of the continent, produced an ice cap several hundred miles wide. The Earth was perpetually locked in the iron grip of winter, blocking the passage north. Present-day Spain was isolated by the frozen barrier of the Pyrenees. From what is now France, a single narrow corridor ran in the direction of Germany and the Danube, between the southernmost limit of the Greenland Ice Sheet and the Alpine Arc. To the centre and east of Europe, the development of huge glaciers on the Caucasus Massif increased the Neanderthals' geographical and geological isolation.

Under the effect of what is known as 'genetic drift', a new type of human appeared within the European ice-trap. Of course, some of his characteristics resulted from a biological adap-

OPPOSITE PAGE AND ABOVE
Neanderthal Man was a form of *H. erectus*. Trapped in Europe, he had been forced to make specific adaptations to the changed climate.

tation to the new conditions. But the permanent acquisition of his unique physical characteristics is doubtless due to the combination of chance, demographic crises and the consequent accentuation of genetic phenomena.

In 1856, two quarrymen discovered, in a cave in the Neander Valley near Düsseldorf, a skull and some bones which revolutionised human palaeontology. At the time, though, the discovery aroused curious reactions; some scientists declared that it was the skull of an idiot, a person deformed by rickets, or even a Cossack deserter from the Napoleonic Wars!

Happily, other similar skeletons were brought to light, notably in a cave near La Chapelle-aux-Saints, during the summer of 1908. All doubts were banished; Neanderthal Man really did exist. What is more, he occupied a unique, but vital, place in the evolution of the genus *Homo*, being the perfect illustration of specific adaptation to unique environmental conditions. He reigned supreme in Europe for tens of thousands of years before other humans took over his mantle.

Where did they come from?

In the opinion of some palaeoanthropologists, the Neanderthals descended from populations of *H. habilis*, or, at a later period, from *H. erectus* who arrived in Europe prior to 1 million years ago. Half a million years later, fossils display the features of 'neanderthalisation', particularly in the nasal and infraorbital areas. These traits are especially pronounced in *H. heidelbergensis* (pre-Neanderthal Man) and even more so in Tautavel Man.

Discovered in 1971 in the eastern Pyrenees by Henry de Lumley, Tautavel Man – dated at 450,000 years – bore a striking resemblance to the Neanderthals. Specialists classify him as a pre-Neanderthal.

The characteristics of the back of the skull are still more marked between 400,000 and 250,000 years ago, for example in fossils found at Swanscombe (England) and Steinheim (Germany).

It was only from 200,000 years ago that the Neanderthals, with a few exceptions, acquired their 'classic' features.

LEFT AND OPPOSITE
A fleshy face, highly prominent nose, a projection above the eyebrows, a voluminous skull . . . In the early twentieth century, Neanderthal Man was considered an under-developed brute.

THE FEARSOME FACE OF NEANDERTHAL MAN

Long considered as a primitive brute, *H. neanderthalensis* was very different from humans today. His brain was astonishingly modern, but his face sends shivers up our spines.

The skull of the so called 'classic' Neanderthal is very characteristic. It displays a whole host of peculiarities, both old and new, without parallel in the prehistoric world – the result of the species' adaptation to climate change.

Among the archaic features are a receding skull with a heavy ridge (supraorbital torus) overhanging the face. The jaw is also receding and the chin non-existent. The Neanderthals' incisors and canines were as powerful as ever, suggesting that they used them as a vice in activities other than eating.

Other features, however, are more reminiscent of modern Man, notably the capacious brain of 1,600 cm^3 (98 cubic inches), which was actually slightly larger than ours. The extraordinary development of the brain is invaluable evidence of Neanderthal Man's adaptation to a harsh climate, though this was never so severe as that of the present Arctic zones. Study of the interior of his skull, with its very thick bony walls, reveals that his brain enjoyed a less efficient blood supply than that of the first *H. sapiens* who evolved on the other side of the ice barrier, particularly in the Near East.

There were specific features distinguishing *H. neanderthalensis* from other representatives of the human race. His skull was elongate, with a rear protrusion resembling a bony nape. His face was very long, with high, rounded eye-sockets and a huge nasal cavity. The cheekbones had completely disappeared and the infra-orbital area slanted markedly backwards. The nose itself was large and highly prominent.

Specialists have been able to reconstruct the Neanderthal face. It is somewhat puffy and the scalp is particularly striking, pulled tautly

backwards by muscles attached to the back of the skull. This gave rise to the earliest descriptions of Neanderthals as unintelligent brutes: a concept that endured into our own times. This was how they were depicted in films like Jean-Jacques Annaud's *La Guerre du feu*, with their furs, wooden clubs and grimacing faces.

The characteristics of Neanderthal Man make clear that he cannot be considered as a primitive form of modern Man; he is a distinct evolutionary branch of the hominids, with a parallel, but separate, development from our own.

Human ... but different.

Conflicting theories abound. When the first Neanderthal was discovered, scientists jumped the gun somewhat, concluding they were dealing with an ancestor of modern Man, archaic, brutish and visibly less intelligent. But new work carried out over recent decades has proved beyond doubt that modern Man was not a linear descendant of the Neanderthals. The two constituted distinct, if contemporary, populations, living in different geographical zones. Thanks to numerous discoveries, from Brittany to Iran, the Neanderthals have become the best-known to science of all prehistoric humans. Such a harvest of fossils (over 200 specimens, adults and children) has enabled us to make detailed studies of the marked differences between their make-up and ours.
In addition, tests using advanced techniques like electron microscopy have yielded very important information about the Neanderthal diet. And remains of fossilised DNA from the specimen recovered in 1856 have led to the unequivocal verdict: *H. neanderthalensis* and *H. sapiens* belonged to different evolutionary branches.

ABOVE AND OPPOSITE The brain capacity of Neanderthal Man was equal, even superior, to that of *H. sapiens*, his contemporary.

A STANDARD TOOLBOX

After more than 1.5 million years of infinitely gradual development, stone tools had now become standardised. Neanderthal Man perfected new methods, displaying at the same time a certain fascination with the abstract notion of beauty.

OPPOSITE
With the advent of the Neanderthals, ancestral scraping implements were complemented by inventions like borers, gouges and denticulate tools, resulting in ever more efficient hunting weapons.

The physical developments of the Neanderthals were only one facet of this gifted race. They also evolved an original culture, based on the stone tools produced by their ancestors. They improved on the preceding Acheulean system and, with *H. sapiens*, originated the civilisation known as Mousterian. Distinctive characteristics of this cultural period are the slimmer, more elegant tools and the precise fashioning of flakes. But the real revolution pioneered by *H. neanderthalensis* lay in preparing flint cores in such a way as to obtain a flake of predetermined and precise form – the 'Levallois technique'.

Making tools of this quality demanded mental faculties far in advance of those enjoyed by the australopithecines. Artisans had to work out the shape of the desired tool, and therefore its purpose, before they could start, as well as planning the various stages of the flaking process. Some would even travel many miles to find flints of the right quality. The Lower and Middle Palaeolithic periods (from 700,000 years ago) have yielded a multitude of artefacts proving that the Neanderthals were intelligent enough to devise ways of working with stone to suit every need. Or, if they did not invent these, they perfected them.

The Neanderthals' methods of stone-working highlight the level of their cultural achievements, in that tools were now standardised among all population groups. The range of scrapers inherited from their ancestors was extended by borers, gouges and denticulate implements. Neanderthal culture is known as Mousterian, after a cave (Le Moustier) in the Dordogne, where the first artefacts were discovered. It is typified by an immense diversity of forms which rubbed shoulders for tens of thousands of years without becoming to any degree hybridised.

Cultural factors played an important part at this period, with different regions developing their own technical and technological solutions from identical raw materials, flint in particular. Regarding the transmission of this quasi-industrial knowledge, Jean Piveteau declares in *The Origin and Destiny of Man* (1983): 'The manufacturing techniques of the Mousterian industries were too complex to be communicated by mere imitation: there must have been teaching – and consequently language.'

The Levallois technique

Some 300,000 years ago a new technique appeared, described for the first time at Levallois, near Paris. It had originated in Africa, but was only transmitted to Europe much later, and persisted until the arrival of the first *H. sapiens* some 40,000 years ago. Instead of chipping successive flakes from a lump of flint like their predecessors, the Neanderthals carefully worked on the flint nucleus. With the aid of a percussion tool made of antler or softwood, they removed fragments from the core surface to form a platform for the predetermined break. Then, with a final blow, they removed a large, thin slice known as the 'Levallois flake'. To become a tool, this only needed light finishing according to its intended use. The tools they made could be used for scraping skins, carving meat, fashioning wooden weapons and many other purposes.

Neanderthal Man had access to a wide variety of tools answering every purpose. This so-called Mousterian culture flourished throughout Europe.

The Neanderthals became a force of nature. They learned to protect themselves by covering their bodies with skins and furs – the first clothing. Neanderthals were the most enduring and the most powerful of all our ancestors.

the carcass. Additionally, when viewed by electron microscopy, flint tools such as scrapers reveal obvious traces of having been used on pelts. And thus the fur trade was invented!

We are now in a position to paint a full portrait of what was in every sense an exceptional being, a pure product of Europe, hardy in the extreme, his territory varying according to the advance and retreat of the ice. The remains collected over the past decades reveal that the Neanderthal populations, their numbers diminishing, regrouped towards the south of the continent as the glaciers advanced. Meanwhile they continued to perfect their stone tools and invented new social relationships. Unique, yes; brutes, never!

BELOW AND OPPOSITE Neanderthals hunted to feed and clothe themselves. But they diversified their diet, sometimes eating fish or small shellfish.

FOLLOWING DOUBLE PAGES Skilled hunters, intelligent and curious, boasting advanced lithic technology and sophisticated language, the Neanderthals reigned supreme in Europe for thousands of years.

The Neanderthal cave

Vestiges of Neanderthal occupation are found almost exclusively in caves or rock shelters.

They chose the site of their caves with great care: it was preferable to have the mouth facing south or west rather than due north, for example, so as to enjoy maximum sunlight. Other factors to consider included the prevailing wind – the cave was first and foremost a shelter – and the nature of the surrounding site. The best position was on a cliff, with a view of the valley below, its game and potential enemies. In practice, few sites or caves fulfilled all these criteria.

Space in the caves was arranged for maximum efficiency. We can still see how areas were reserved for hearths or toolmaking, the former recognisable from traces of fire or the remains of butchered animals. To all intents and purposes, caves were the forerunner of the modern house.

HOMO SAPIENS

IN OUR OWN IMAGE

Whether his origins lay in Africa or the Middle East, a new actor appeared upon the stage of human history. His technology and spirituality earned him the name *Homo sapiens*: the Wise One. And he would inherit the Earth.

THE WORLD THEIR KINGDOM

Between 100,000 and 30,000 years ago, *H. sapiens* migrated throughout the planet, travelling as far as Australia and over the Bering Straits, then a land bridge, to America.

Some 40,000 years ago, a new type of human being appeared on the borders of Europe before rapidly colonising the rest of the globe. The origins of *H. sapiens* are still relatively unclear, at least until 100,000 years ago. We know that he already existed at the time in the Near East, from where he must therefore have travelled to western Eurasia. But, for the moment, his exact whereabouts much before his arrival in the Near East must remain a mystery.

There are two principal theses: the 'Noah's Ark' hypothesis (single point of origin) and the 'candelabra' theory that proposes several departure points.

For the partisans of the first, the origin of modern Man is unique, and, by extension, African. Beginning with the great original continent, he progressively overran the whole of the ancient world via the Near East. He then headed in one direction towards Europe, and in the other to Asia, slowly but surely replacing the

OPPOSITE AND RIGHT With his ability to walk long distances, his social organisation and inborn inquisitiveness, *H. sapiens* swiftly colonised the planet.

older populations of these regions like the Neanderthals of Western Europe.

The second theory, also known as multi-regional origin, claims that modern Man arose as a direct result of evolution in several parts of the ancient world (Africa, Asia, Near East) from more primitive populations, themselves descended from *H. erectus*. This argument is based on the discovery of fossils revealing a certain continuity, right down to modern Man in fact, in the regions concerned.

So far, however, the factual evidence does not really add up. Since the 1930s, numerous fossil finds – particularly in various parts of Africa and Israel – have constantly revived the controversies surrounding the origin of these archaic types of *H. sapiens* who supposedly formed the second great wave of prehistoric emigration.

In 1997, the American palaeoanthropologist Tim White and his Ethiopian colleagues discovered hundreds of hominid bone fragments in the middle valley of the Awash. When pieced together, they formed three skulls, including a child's or adolescent's, dating to 160,000 years ago – broadly speaking, the Neanderthal period. Their traits, resembling those of modern *H. sapiens*, add weight to the African hypothesis. Other researchers, however, such as Marie-Antoinette de Lumley, openly question the modalities involved in the replacement of archaic peoples by 'true' men in a second wave from Africa.

A decisive judgement, then, is still in the future. Nonetheless, the fact remains that in less than 100,000 years *H. sapiens* came to dominate the whole of the ancient world before launching an assault on America via the Bering Straits. No conditions were hostile enough to halt his progress.

Cro-Magnon: Universal Man

***Homo sapiens* is often described as Cro Magnon Man,** especially in the case of European populations. However, use of the name has been extended to all human species of the Upper Palaeolithic.

In 1868, at Eyzies-de-Tayac in the Dordogne, in a rock-cavity on the property of a Monsieur Magnon, there came to light the most famous of Europe's *H. sapiens* fossils. In the local dialect, 'cro' means 'hole', so 'Cro-Magnon' could be translated as 'Magnon's hole'.

Excavation methods in those days were much more rudimentary and lacked the scientific precision to which we are accustomed. Palaeoanthropologists do not, unfortunately, possess any data that might help identify the populations or at least date the fossils. In addition, remains are often so slight (two teeth for the Uluzzian period) that it is virtually impossible to draw any substantial conclusion about their owners.

LEFT AND OPPOSITE
H. sapiens made a great technological leap forward. Besides excelling in the manufacture of stone tools, he invented a multitude of new implements made of wood and bone.

A TECHNOLOGICAL LEAP

They used strange weapons. They also caught fish, which they dried like meat. They carried spears with fine, pointed heads. Using small, claw-like bone 'needles', they were able to sew furs and skins.

The Upper Palaeolithic, from 40,000 years ago, was the beginning of a period dominated by innovation, either technical or cultural. As time progressed, *H. sapiens* turned to hitherto unused raw materials like wood, the antlers of cervids (including reindeer) and various types of bone, devising new forms of tools for new purposes. Among these were a harpoon for more efficient fishing – and sewing needles with eyes, leading to the earliest experiments with 'made-to-measure' garments! He fashioned jewellery, necklaces and bracelets. His casting weapons became more deadly; in particular, the invention of the atlatl or spear-thrower, during the Magdalenian period (from 13,000 years ago), meant that dangerous game could be brought down with less risk to the hunter. He also developed new tracking techniques and could now complement his traditional diet by fishing, enjoying a wider range of proteins as a result.

With this progress went hand-in-hand an improvement in flint-knapping. In the Solutrean period (27,000 to 13,000 years ago) the technique reached a virtual summit of perfection in the large 'laurel leaf' points only a few millimetres thick, revealing a consummate skill on the part of the artisans.

Lifestyles also diversified. *H. sapiens* established himself in every sort of habitat – caves, by preference, but also tents and huts constructed in the open. Evidence of the latter comes, for instance, from the post-holes found at the Pincevent site (Seine-et-Marne, France), discovered in 1964 and dated to 8,000 years ago.

To provide himself with food, clothing, shelter, weapons and tools, *H. sapiens* exploited every available resource and raw material from his environment.

To make these shelters, *H. sapiens* used every resource of the environment. In cold regions like the Ukraine he constructed his dwellings from mammoth bones with a covering of skins, while at Pincevent reindeer hunted locally provided essentials like food, clothing and tenting.

By now, Man was a past master at manipulating fire, and could keep the hearth at just the right temperature for his needs. He employed different fuels for light, heat or cooking: wood, bones, peat, etc. The Pincevent site, for instance, reveals an alignment of three hearths surrounded by debris such as reindeer bones or 'debitage' – discarded flakes from the manufacture of lithic tools. Eight other encampments have been discovered in the vicinity, indicating that the site was highly prized at the time for its natural resources.

Mankind was, at this epoch, divided into small groups, nomadic or semi-nomadic. The wanderings of these groups were often seasonal as they followed the game herds or returned to cave shelters for refuge from the winter.

BELOW AND OPPOSITE The invention of bone tools (osteodontokeratic culture), notably sewing 'needles' with eyes, greatly improved living conditions. Men were able to make warmer, more sophisticated clothes and tents to ward off the harsh climate. They also devised new techniques for preserving food, such as drying and smoking.

The beginnings of international trade

The Earth's human population was highly dispersed during the Upper Palaeolithic period. Groups of hunter-gatherers met only rarely, yet we can imagine that each encounter led to a wealth of exchanges, both cultural and technological. An embryonic form of international trade has been discovered at virtually all sites of the era. Each yields its store of surprises, from lumps of raw materials to weapons and jewellery. Always that little something turns up where least expected, like a flint nucleus from a quarry sometimes hundreds of miles away or a fragment of amber acquired from some settlement far from the natural deposits of this ancient and magical substance.

These new breeds of men were travellers, keeping themselves up to date with the development of technology, exchanging tools and ideas. Just like we do today.

OPPOSITE
The earliest European form of *H. sapiens*, Cro-Magnon Man, was our ancestor. He resembled us in many ways both in appearance and culturally – not surprisingly, as we are the same species.

DOUBLE PAGES PRECEDING AND FOLLOWING
With a brain that had become exceedingly complex, *H. sapiens* ('Wise Man') turned his attention to his surroundings and their significance.

A HANDSOME SPECIMEN

With his highly 'gracile' physique, 'anatomically modern Man' was very different from his Neanderthal cousins and the most streamlined of all human beings to date.

When he arrived in Europe some 40,000 years ago, 'anatomically modern Man', as he is described scientifically, was very different from the native Neanderthals.

Firstly, the newcomer had a unique skull. Its general form was rounded, with a vast cranial capacity – averaging around 1,450 cm^3 (88 cubic inches) but sometimes as large as 2,000 cm^3 (122 cubic inches). The crown was raised and broader at the highest point; most or all of the jawbone (now smaller) was sited perpendicularly below the braincase. Moreover, these were the first men to have a proper chin.

Generally speaking, the cranial bones were thinner and finer with less prominent projections. For example, the ancestral ridge above the eyes (supraorbital torus) was substantially reduced, even if some skulls of the period – and our own – preserve signs of it to varying degrees.

It is noteworthy that these evolutionary developments between the skulls of archaic and modern *H. sapiens* took place relatively rapidly, that is, over only 100,000 years. The skeleton of *H. sapiens* also acquired a distinctly more gracile form, especially in regard to the legs, proportionally longer and less bandy. *H. sapiens* walked with the flat of his foot on the ground, and hence more efficiently. His arms were also longer in proportion. This streamlining gave him better balance, as the centre of gravity of his body was now measurably closer to the axis of his spine.

DNA to the rescue

Thanks to advances in molecular biology, it has become possible to extract and amplify DNA trapped in fossil remains. Unfortunately, well-preserved samples are rare, as all organic matter rapidly degrades after death. This explains why the dried skin of individuals who have been mummified or imprisoned in ice for centuries represents the best chance of finding DNA in reasonably good condition. But periods as remote as these, the likelihood of finding such samples is very small.

In 1997, a team from Munich University succeeded for the first time in extracting and studying DNA from a Neanderthal who died at least 30,000 years ago. Since then, other specimens have yielded up a few molecules, allowing researchers to carry out more in-depth investigation into the variability of the species. Obviously, these DNA fragments are minute: around 350 nucleotides, whereas the present human genome contains between 3 and 3.5 billion. Yet the differences observable from these fragments are apparently large enough to prove that Neanderthals were a species distinct from modern Man.

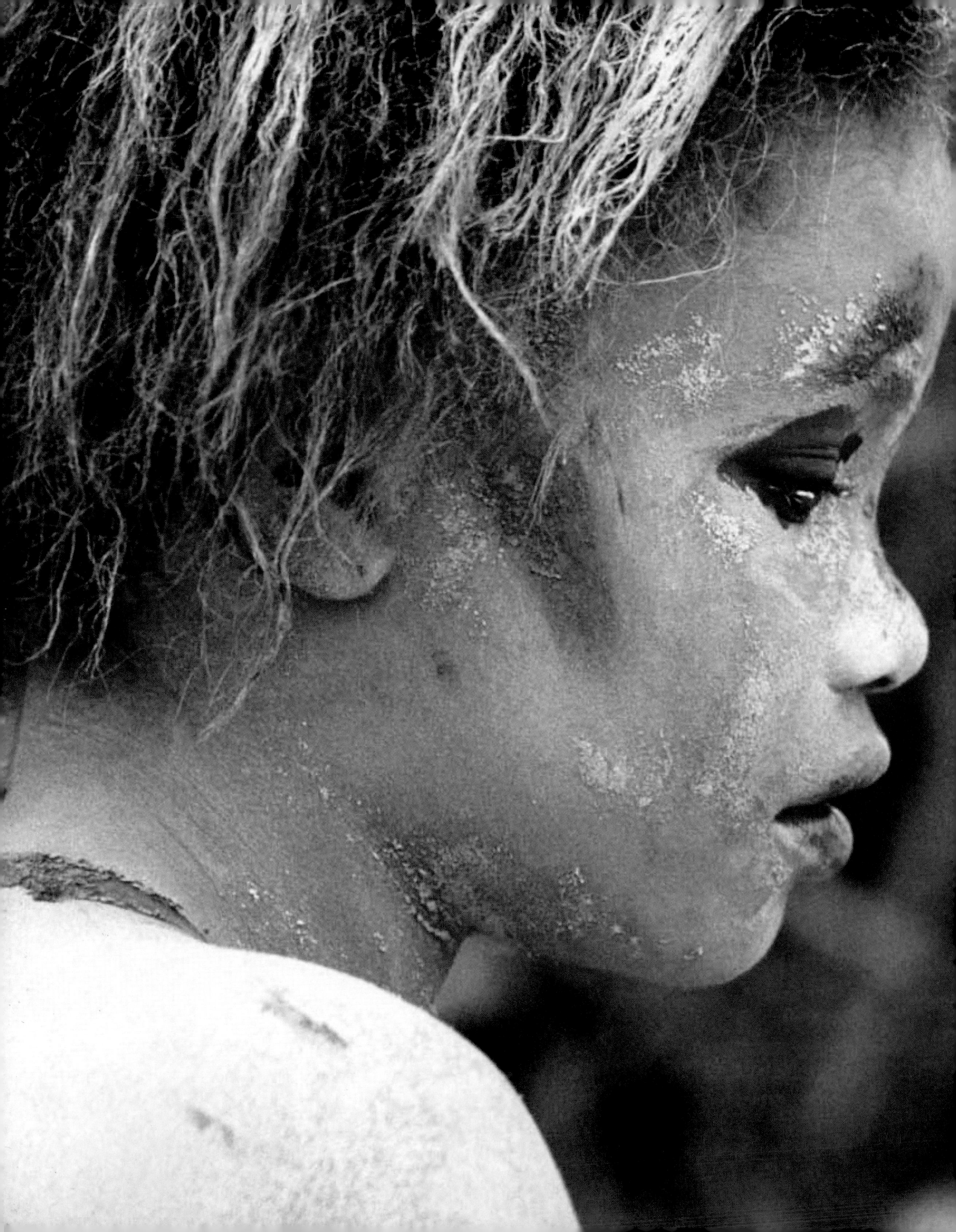

A CLASH OF CULTURES

Before becoming extinct, Neanderthal Man came face to face with *H. sapiens*. He must have been intrigued by his cousin's appearance, with his body-ornaments, jewellery and his newfangled tools and weapons.

ABOVE AND OPPOSITE
Differences in language and culture must have made communication between Neanderthals and *H. sapiens* extremely difficult, though the two human species made limited exchanges of ideas and technology.

The Neanderthal culture was entering its last few millennia when the first groups of *H. sapiens* set foot in Europe. The two species must have coexisted in what was in fact a fairly restricted area, though positive proof of this is still wanting. But the nomadic nature of the two civilisations would certainly have made it possible for them to meet and make exchanges.

The Châtelperronian era (37,000 to 30,000 years ago) takes its name from the site at Châtelperron (Allier, France) where its toolkit was first identified by Abbé Breuil in 1906. This so-called transitional period gives us a better insight into the mingling of cultures which accompanied the replacement in Western Europe of Neanderthal populations by 'anatomically modern Man'. Characteristic tools are scrapers, burins, backed blades, punches and bevelled spear points, with additional items that 'drifted' in from the Mousterian period. Evidence from several sites attests to the use of 'soft' striking tools, made notably of bone. One of the most advanced developments was a small flint knife whose blade, with its curved back, was formed by repeated gentle chipping.

It is a common if somewhat controversial assertion that tool manufacture of this sort was a Neanderthal development, just as modern Man created the contemporary industry of the Aurignacian period (38,000 to 26,000 years ago). Existing side by side, both industries produced equivalent items, making it very likely there was an exchange of technologies.

The Neanderthals rubbed shoulders with *H. sapiens* for thousands of years. There is no evidence of wars or genocide. But they were, it seems, unable to unite to breed a new human race.

BELOW AND OPPOSITE
Neanderthal Man probably engaged in limited cultural exchange with *H. sapiens*, his contemporary. But he would still have been bemused by his neighbours' tools, such was their diversity and refinement.

DOUBLE PAGE OVERLEAF
The first musical instruments are attributed to the Neanderthals. But it was *H. sapiens* who perfected the arts of making and playing them.

Moreover, in July 1976, the cave at Saint-Césaire (Charente-Maritime) yielded up the partial skeleton of a Neanderthal surrounded by tools of an obviously more advanced manufacture than those of the Châtelperronian or Mousterian. It appears that the local Neanderthals had adopted techniques introduced by *H. sapiens*. Other evidence – from the site at Arcy-sur-Cure (Yonne) in particular – reveals that both *H. sapiens* and Neanderthals employed ivory, but with different techniques, to make a type of pierced jewel.

The debate continues as to exactly how these exchanges of technology occurred. 'Cultural barriers are always harder to cross than geographical ones,' in the words of Henry de Lumley. Nonetheless, both *H. sapiens* and *H. neanderthalensis*, contemporaries in Western Europe, practised the same funerary rites, used equivalent tools and weapons, and were both acquainted with music. Did *H. sapiens* teach the Neanderthals, or vice versa? Cultural exchanges hardly took place on a one-for-one basis, but we await new evidence of the true scale involved.

The music of our ancestors

The first known musical instruments date back 45,000 or even 60,000 years in Central Europe, and can be attributed to the Neanderthals. Many others, belonging to *H. sapiens*, have also been discovered. Our ancestors were acquainted with the musical bow and employed large skulls as percussion instruments, the latter still bearing the marks where they were struck. Investigators have also discovered, in Ukraine, a sort of orchestra composed of wind, string and percussion instruments! On the other hand, it is to all intents and purposes impossible to access the actual songs, music and dances of the period – the only artistic legacy of this period is the carvings and paintings found in caves.

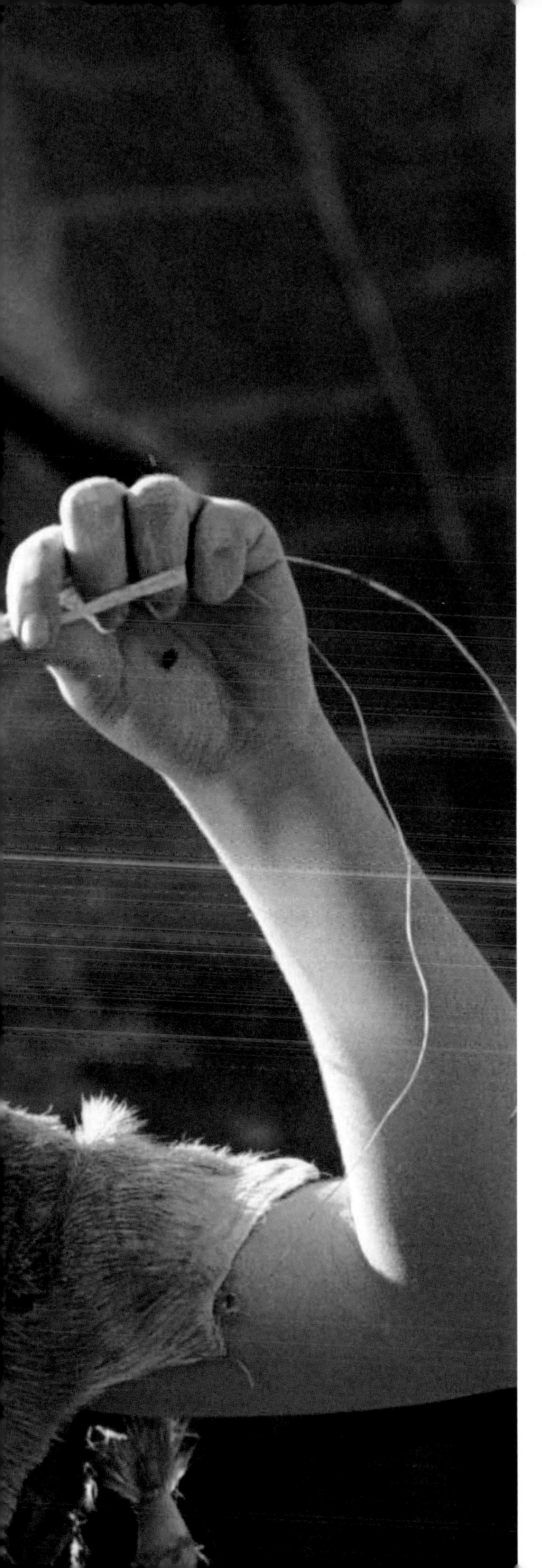

LEFT
Personal ornaments such as necklaces and bracelets were among the many means of artistic expression developed by *H. sapiens*.

ART AND NATURE

The birth of art is important evidence of the capacity of modern Man for analysis and communication, to think in terms of the abstract and the ideal.

Around 50,000 to 45,000 years ago human civilisation took a giant leap forward in its capacity to develop complex ideas by learning how to express the universe in symbolic fashion. The means of representation Man devised – cave painting, rock-carving, sculpture, metal-casting, jewellery, funerary furnishings – would become the foundation of the visual and plastic arts.

This artistic explosion took place throughout the world simultaneously, leading some experts, like Jean Clottes, to assert that a capacity for artistic expression is a quintessential feature of Man, encrypted in his genetic make-up.

Prehistoric art falls into a number of categories including *art mobilier* or 'portable art' (statuettes, plaquettes, figurines) and the wall art found in caves, both of which evolved over time. The earliest examples date back to the Aurignacian period and belong predominantly to the first group. Female figurines, their lines flatteringly exaggerated, appeared throughout

H. sapiens crossed the oceans and migrated throughout the planet. In parallel with this global conquest he evolved art forms displaying astonishing skill from the very beginning.

Europe between 25,000 and 20,000 years ago. The Hohlenstein-Stadel site in Germany produced a statuette known as the Hohlenstein Lion-Man, symbolism at its most powerful and obvious.

By universal agreement the most spectacular of all early art forms is cave painting. By 32,000 years ago it had already matured into the exquisite imagery found in the Chauvet Cave at Vallon-Pont-d'Arc in the Ardèche or, a little later, at Arcy-sur-Cure. Men of the Aurignacian period continued to develop this highly sophisticated form of expression until it culminated, during the Magdalenian period, with the Lascaux Cave in the Dordogne – the 'prehistoric Sistine Chapel'. At this one site, 17,000 years old, researchers have counted at least 600 paintings and 1,500 carvings.

The Epipalaeolithic art of these advanced hunters exhibits a great diversity of style. Rock paintings very frequently depict hunting scenes with large herbivorous mammals such as horses, mammoths, bison, rhinos or reindeer. Yet they ignore every feature of the natural surroundings: there are no trees, no mountains, stars or sun. The paintings are not meant to be representations of the real world; rather, they follow a strict pattern of codes and conventions. Another interesting feature of cave art is that close examination has, in some cases, revealed fine lines indicating the use of brushes made of animal hair.

Unfortunately, many of the mediums involved, being of plant origin (e.g. wood, bark), have not resisted the ravages of time and are lost to modern study. But ongoing inventories have already catalogued over 30 million examples of cave paintings and carvings and 100,000 *objets d'art* in 150 countries worldwide. The sites know no geographical bounds; they include the Tassili Plateau in the Sahara, the rock-shelters of the Drakensberg in South Africa, Mont Bego and the Valley of Wonders (Alpes-de-Haute-Provence), the Indonesian forests and the Utah canyons.

BELOW, OPPOSITE, AND FOLLOWING PAGES
Cro-Magnon Man left many traces of contemporary art forms, particularly in his caves. But others, like body-painting, singing and dancing, are lost to us for ever.

Dating cave art

It is relatively easy to date the portable art of the Palaeolithic period from the layers in which the objects are found. Wall-paintings, carvings and bas-reliefs are a different matter, and defied precise dating until recent scientific advances. Earlier specialists like Abbé Breuil or André Leroi-Gourhan based their estimates on stylistic factors. For several decades the carbon-14 method was used to date specimens of paint, pigments and charcoal. The arrival of the particle accelerator meant it was necessary to sample only a few milligrams of pigments from the paintings, whereas early radiochronology techniques required several grams. These new methods of dating cave art have produced some surprises, sometimes revealing that our ancestors returned at different periods to work on the same site – 32,000 and 18,000 years ago, for instance, in the Chauvet Cave. Clearly the memory of this unique and magical location was handed down over countless generations.

RIGHT AND OPPOSITE Comparisons with certain modern societies suggest that Cro-Magnon Man engaged in shamanic rituals, particularly for funerary rites.

FROM ART TO RELIGION

Our near ancestors buried themselves in the darkness of caves to make their paintings. Why? Was religious ritual behind their art?

The human form is rarely seen in cave art. When it does appear, it is highly stylised, clashing with the sometimes almost photographic qualities of the large mammals. When cave art was first discovered in the late nineteenth century, specialists in prehistory saw it as merely a means of self-expression: art for art's sake. But it was not long before deeper meanings began to emerge.

Most paintings are, in fact, sited in inaccessible locations such as caves, well away from prying eyes. Abbé Breuil, in the 1950s, interpreted these places as magical or religious sanctuaries associated with secret rites. A more recent hypothesis is that they were linked to shamanic practices.

Some of the non-figurative works continue to astonish researchers. Points, lines, geometric shapes – meaningless to us – might, according to André Leroi-Gourhan – be 'mythograms' (i.e. systematic representations of ideas). They are quite different to hieroglyphs and as yet indecipherable, but they appear to have some precise significance. Was this the beginning of a written language, or perhaps a non-verbal one (enshrining gestures, body-language) connected with hunting? Probably we shall never know.

Comparisons between various sites reveal traces of what appears to be a universal pictorial language, based on a codified system associated particularly with animals. But modern science has yet made little of this highly elaborate pooling of ancestral lore.

The cave art tradition persisted for over 25,000 years both in Europe and in sites beyond its boundaries. Such timescales suggest that there must have been a very strong link between populations – a structured and defined means of communicating knowledge, myths and representations of the world, based on some form of religion.

One of Man's innate characteristics is spirituality. He could visualise an afterlife, ceremonially daubed cadavers with red ochre (the

BELOW, OPPOSITE AND DOUBLE PAGE OVERLEAF
The first *H. sapiens* buried their dead according to very precise rituals and with the assistance of a shaman whose duty it was to open the door to the world beyond. Whether they borrowed these customs from their Neanderthal cousins is uncertain.

PAGES 162–163
The last Neanderthals disappeared 25,000 years ago. *H. sapiens* became the sole surviving human. But for how long?

colour of blood, i.e. life) and arrayed them with jewels, rather than leave them to the mercy of carnivores. He was able to portray symbolically the relations between men and animals and idealise those of man and woman. These activities reveal mental and psychical processes unique in the story of life on Earth.

We still have an immense amount of work to do to understand this link between the material and the spiritual worlds. We have a duty to our ancestors to make the effort. After all, we are only here now because they pioneered the way …

The human hand: a universal symbol

Everywhere art exists, men, women and children have left their hand-prints for posterity. Depending on the technique, these prints appear as positives or negatives. For negative prints, a method is used akin to stencilling; the artist blows a mixture of water and pigments (ochre, manganese, for instance) from his mouth around his hand, which thus appears in outline. Positive prints are made by simply smearing the hands with paint and pressing them against the rock. Two-thirds of prints discovered show missing fingers. Some experts see this as a sign of voluntary mutilation: an initiatory ritual or mourning rites. Others, like André Leroi-Gourhan, have put forward a more probable explanation, given that these men lived by hunting and no skeleton has so far been discovered with missing fingers. He believes the prints were made with the fingers folded back, a coded gesture linked to hunting rituals.

CHRONOLOGY OF DISCOVERIES

1771 First recorded discovery of prehistoric human remains by Johann Friedrich Esper in Germany.

1823 William Buckland discovers 25,000-year-old skeleton bearing traces of red ochre and surrounded by ivory jewels – the 'Red Lady of Paviland'.

1848 Well-preserved adult skull unearthed on Gibraltar; later regarded as belonging to relative of Neanderthals.

1856 Discovery in Neander Valley, near Düsseldorf, Germany, of Neanderthal man. Baptised *Homo neanderthalensis* in 1864.

1868 Discovery in the Dordogne of most famous *Homo sapiens* fossils: Cro-Magnon Man.

1892 In Indonesia, Eugène Dubois finds a femur proving that fossilised hominids were originally bipedal: *Pithecanthropus erectus*, now *Homo erectus*.

1907 Jaw of human type unearthed near Heidelberg, Germany. Dated to 600,000 years ago; archetypal Heidelberg Man (*Homo heidelbergensis*).

1908 First undisputed Neanderthal burial discovered at La Chapelle-aux-Saints (Dordogne) by brothers Amédée and Jean Bouyssonie leads to re-appraisal of Neanderthal culture.

1925 Raymond Dart discovers 'Taung Child' (*Australopithecus africanus*) in South Africa: the first known australopithecine.

1927 Finding in China of three teeth, lower jaw and skullcap: Peking Man (*Sinanthropus pekinensis*), 500,000 years old.

1936 | 1938 Discovery in South Africa of skull belonging to new species of quasi-human aspect: *Australopithecus robustus*, 2 million years old.

1937 Discovery in Java of 500,000-year-old skull similar to that of Dubois' *Pithecanthropus*. Java Man, Peking Man and Heidelberg Man classified from 1950 as one species: *Homo erectus*.

1940 Children stumble upon Lascaux Cave ('Prehistoric Sistine Chapel'), with 17,000-year-old history and containing 600 paintings and 1,500 carvings.

1959 Mary Leakey's discovery of *Australopithecus boisei* (lived 1.75 million years ago) at Olduvai, Tanzania

1960 | 1964 Discovery at Olduvai by Mary and Louis Leakey of human fossil and worked stone tools belonging to *Homo habilis*.

1967 *Australopithecus aethiopicus* (2.5 million years old) discovered in the Omo valley by Camille Arambourg and Yves Coppens.

1971 Discovery in the eastern Pyrenees of Tautavel Man from 450,000 yeas ago by team under Henry de Lumley.

1974 The most famous australopithecine, 'Lucy', who lived 3.15 million years ago, brought to light by Yves Coppens, Donald Johanson, Maurice Taïeb and their team in the Afar Triangle, Ethiopia.

1984 Discovery at Atapuerca (Spain) of Sima de los Huesos site – shaft containing over 30 skeletons more than 300,000 years old (pre-Neanderthal). The reason for their presence remains a mystery.

1985 Finding at Lake Turkana, Kenya, of substantially complete skeleton of 'Turkana Boy', specimen of *Homo ergaster* from 1.6 million years ago.

1985 Discovery of undersea Cosquer Cave (22,000 years old).

1994 Chauvet Cave (32,000 years old) discovered in the Ardèche by Jean-Marie Chauvet.

1995 1995 Discovery at Kanapoi, near Lake Turkana (Kenya) of *Australopithecus anamensis* (4.2 to 3.9 million years old).
Also, in Chad, by group led by Michel Brunet, of 'Abel', specimen of *A. bahrelghazali* (3.5 to 3 million years old).

1999 Discovery of an *A. garhi* skull by Berhane Asfaw at Hata, Ethiopia. Also two hominid skulls found at Dmanisi, Georgia: baptised *H. georgicus*, but actually specimens of *H. ergaster* from 1.7 million years ago.

2000 Discovery in Kenya of bipedal hominid fossils – *Orrorin tugenensis* – popularly known as Orrorin, by Brigitte Senut and Martin Pickford.
Also found, mandible associated with earlier Dmanisi skulls.

2001 Chad: Michel Brunet and team unearth 'Toumai' – *Sahelanthropus tchadensis*, the oldest hominid (7 million years) known to date.
Also discovery by Meave Leakey in Kenya of skull of *Kenyanthropus platyops* – contemporary of australopithecines 3.5 million years ago.

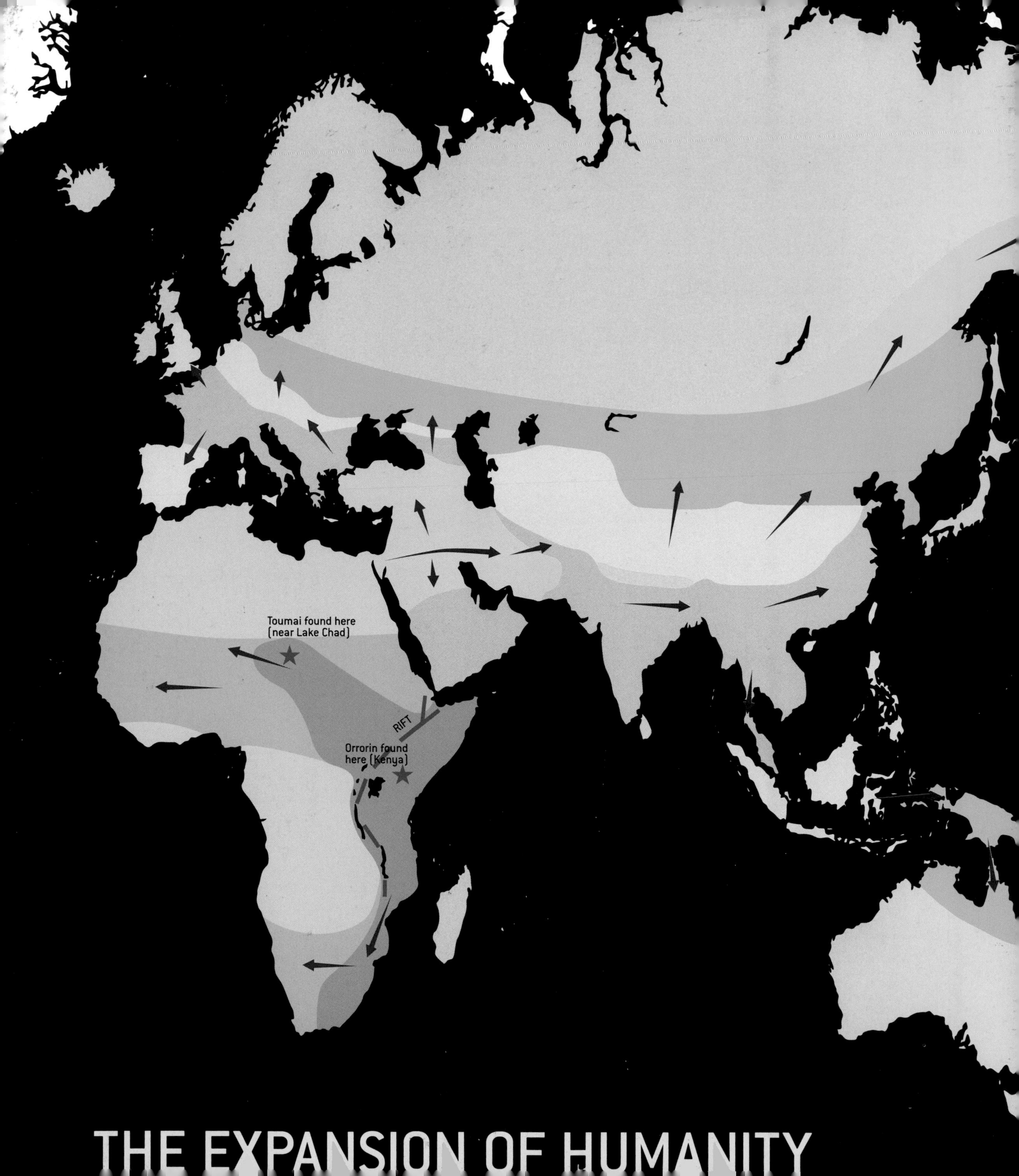
Toumai found here (near Lake Chad)
RIFT
Orrorin found here (Kenya)
THE EXPANSION OF HUMANITY

Bering Strait
formerly land bridge
FIRST PRE-HUMAN POPULATIONS:
7–6 MILLION YEARS AGO
START OF SPREAD OF AUSTRALOPITHECINES:
c. 4 MILLION YEARS AGO
TERRITORY OF AUSTRALOPITHECINES:
4–3 MILLION YEARS AGO
HOMO ERGASTER: c. 2 MILLION YEARS AGO
HOMO ERGASTER AND HOMO ERECTUS:
1.8 MILLION YEARS AGO
HOMO ERGASTER AND HOMO ERECTUS:
1.5 MILLION–100,000 YEARS AGO
EURASIA OCCUPIED; OCCUPATION OF AUSTRALIA:
100,000–50,000 YEARS AGO
OCCUPATION OF AMERICA BY HOMO SAPIENS:
60,000–10,000 YEARS AGO
EXPANSION OF HOMO SAPIENS:
10,000 YEARS AGO TO PRESENT

THE FILM

To re-create a story lost in the mists of time, the crew of *Odyssée de l'espèce* used all the latest techniques, including image synthesis and motion capture.

BEHIND THE SCENES

How were we to re-create prehistory and keep it authentic? How to bring to life characters like Orrorin, Toumai or Lucy who lived millions of years ago? This was the challenge we faced during the making of *Odyssée de l'espèce* and which eventually produced the illustrations featured in this book.

The camera and the computer

There is no way one can use real-life actors to re-create scenes from the very remote past. The physique, expressions and gestures of our ancient ancestors were so different from ours that you could not play them convincingly. Lucy, for instance, was only 1.20 m tall (under 4 ft), with very long arms and short legs. When she walked, she did so with an odd, swaying motion owing to the way her skeleton worked – impossible to replicate. On the other hand, *Homo habilis* and his descendants were beginning to look like us; so, with some crafty make-up and the use of prostheses, it became possible to transform actors into humans of that period.

The result was that we adopted a twofold approach. In the parts of the book dealing with the first bipeds and the australopithecines, and in the opening sequences of the film version, the characters are virtual and superimposed on filmed backgrounds. For the rest, they are played by flesh-and-blood actors and shot using normal techniques.

Virtual stars

Orrorin, Toumai, Lucy and the australopithecines, then, were computer-generated. To begin with, technicians created them as detailed three-dimensional sculptures, video images of which were then fed into the computer where they became 'virtual' characters.

Next, a team of actors played out the scenes in various locations in South Africa, allowing the director and crew to visualise each sequence in its entirety. Footage of the locations was also shot without the actors to produce a backdrop for the virtual characters. At the third stage, the

'The 3D figures are really powerful; you imagine what they'll look like and little by little, you see them come to life.'

OPPOSITE
Actors play the scene wearing devices that capture their movements and expressions.

BELOW AND DOUBLE PAGE OVERLEAF
The virtual characters are animated, then 'clothed' with flesh and hair.

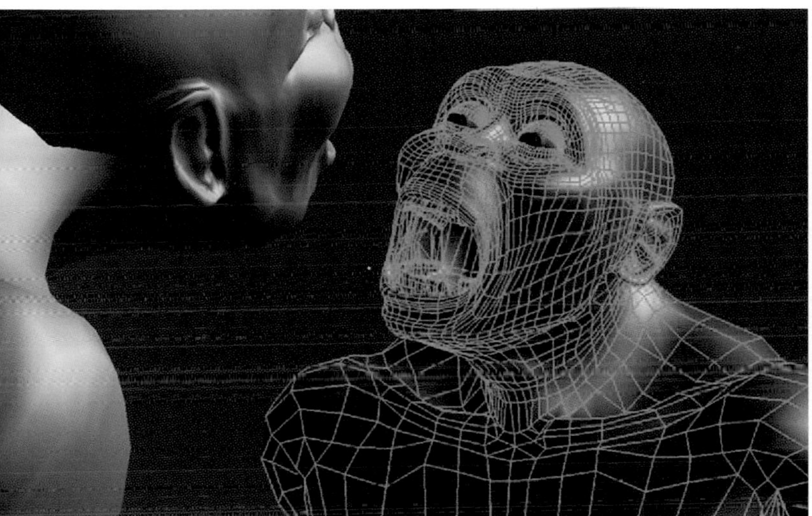

scenes were re-shot in the studio with the cast wearing devices that record movement and, for close-ups, facial expressions. When decoded and analysed, the recorded patterns served as the basis for animating the virtual characters. Finally, the animated figures were electronically provided with skin and hair and slotted into the pre-filmed backgrounds.

A team of 15 graphic artists and engineers from the Mac Guff Ligne studio worked on the project for nine months drawing on all their skills and inventing new digital techniques as they went along to bring our ancient ancestors back to life. Their skin texture, facial expressions and savage looks, scars, fur and hair sculpted by the wind, even their interactions with the environment – no tiny detail was overlooked in the attempt to infuse these synthesised characters with animal vitality and with emotions the audience could share.

Keeping it accurate

Under Yves Coppens' supervision, every stage of the production was monitored by experts to ensure the details were consistent with up-to-the-minute research. Creation of the virtual figures – gait, gestures, 3D processes, etc. – was the province of Anne-Marie Bacon (anthropologist at the CNRS), whose special area of expertise was the skeletons and body movements, and Sandrine Plat (anthropologist, Collège de France) worked on the skulls and facial expressions.

And then there was the real-life cast. Francis Thackeray (Transvaal Museum) was in charge of the preparatory work and shooting in South Africa (sets, costumes, accessories, make-up), while Michael Bisson (Montreal University) oversaw the creation of the masks and prostheses.
There was also the question of the fictional passages, where we had to invent while staying as true as possible to known scientific facts. No-one can be sure, for instance, how our ancient ancestors communicated. We do know, though, that beginning with *H. habilis*, the human mouth possessed a palate, enabling the utterance of syllables. So the writers came up with a type of language corresponding to the probable physical and psychological capacities of the characters we were portraying.

> 'It was wonderful to see all these experts bringing to life beings they had previously only dreamed about or imagined.'
>
> JACQUES MALATERRE (DIRECTOR)

> 'I'd say the film was a work of fiction all right, because it tells stories; but at the same time, it's science.'
>
> YVES COPPENS

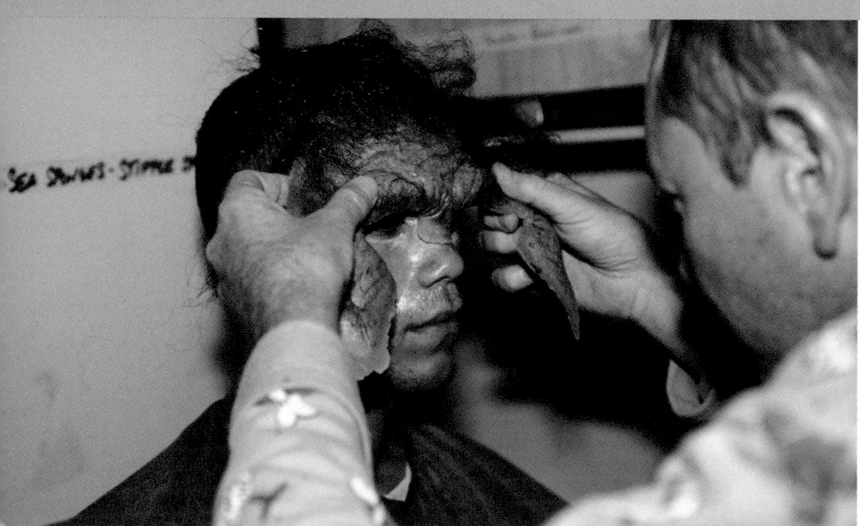

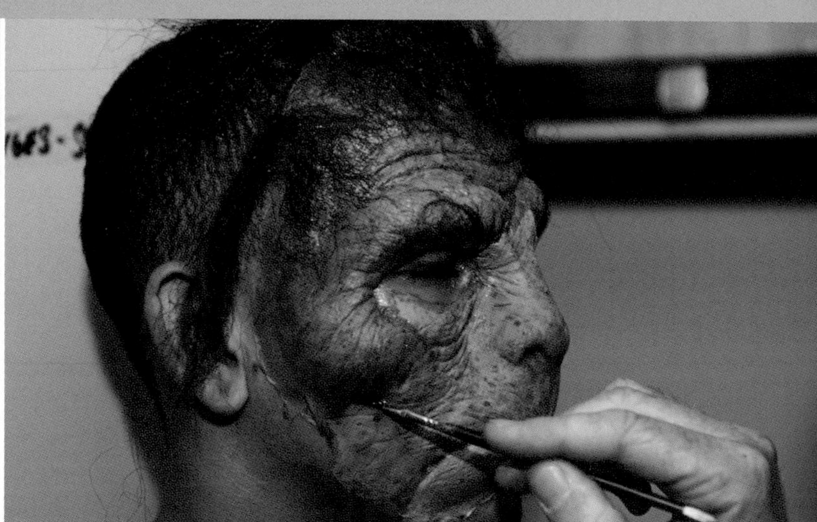

Real-life heroes

Transforming the cast into Early Man demanded an intense amount of work on the part of make-up artists, dressers, etc., under the supervision of recognised experts. No fewer than 45 actors were needed to retrace the adventures of *H. habilis*, *H. erectus*, the Neanderthals, Cro-Magnons and *H. sapiens*, with their masks and prostheses taking three months to make. Each actor had to undergo four hours' make-up every day before going on set. Three separate sections of latex were fastened to their faces with special adhesives, and the expressions painted on to the latex. Finally, body-hair was added. This was gruelling work in temperatures of 30 to 40°C (86 to 104°F). But when Professor Thackeray met the actor playing *H. habilis*, made up according to his instructions, he telephoned Professor Tobias (who had discovered the real thing). 'I've gone one better than you,' he quipped. 'I've just shaken *H. habilis* by the hand.'

ABOVE AND OPPOSITE
Each actor underwent gruelling make-up sessions lasting at least four hours daily.

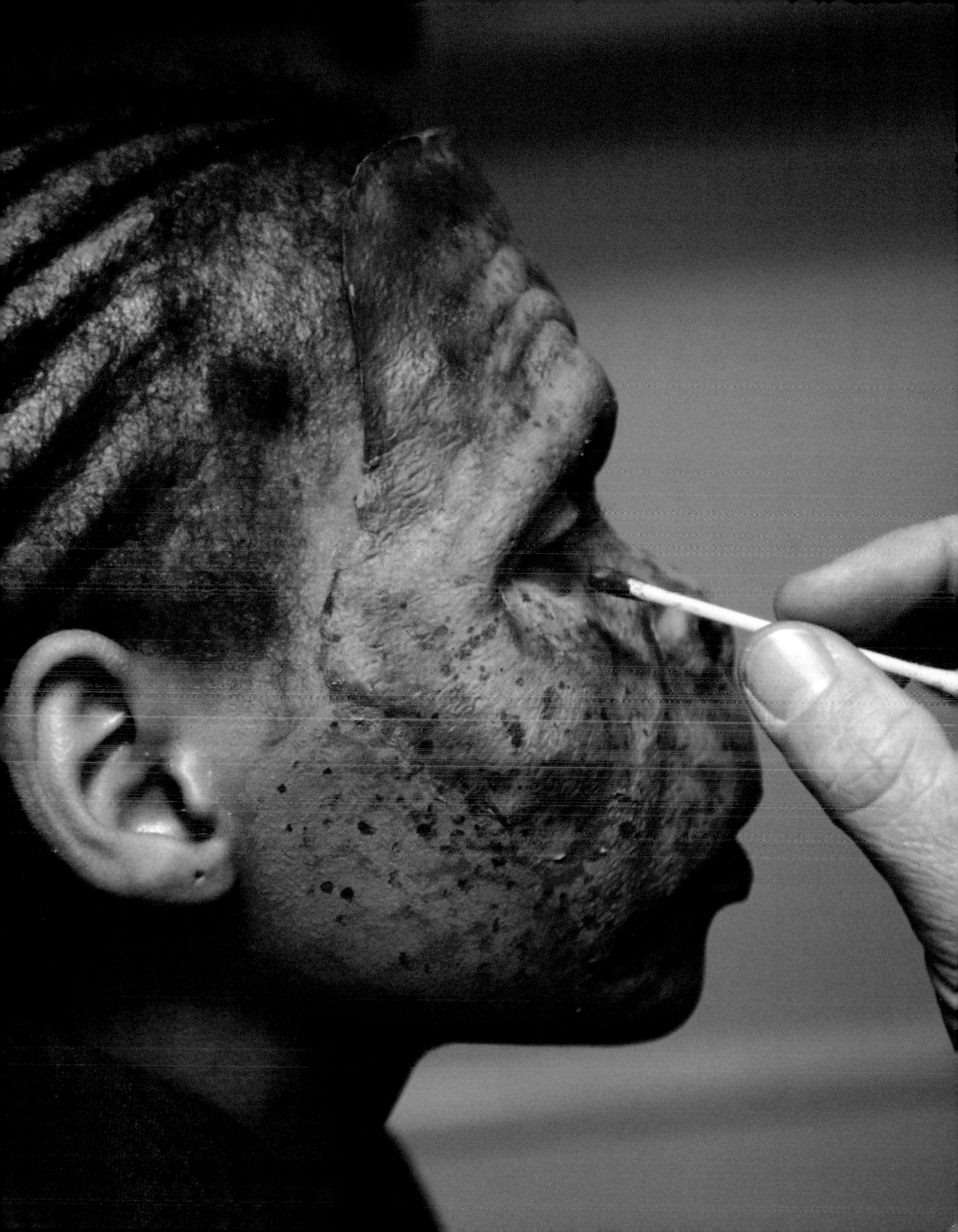

THE FILM

In association with Société Radio-Canada, Discovery Channel Canada, Channel 4, RAI – Radiotelevisione Italiana – RAI 3, ZDF/ZDF Enterprises, Télévision Suisse Romande, Planète, France 5.

Director: Jacques Malaterre
Producers: Charles Gazelle and Christian Gerin
Scientific director: Yves Coppens
Director for France 3 documentaries: Patricia Boutinard-Rouelle
Scriptwriters for 90-minute version: Jacques Dubuisson and Michel Fessler, from an original idea by Hervé Dresen.
Scriptwriters for 3 x 52-minute series: Hervé Dresen, Jacques Dubuisson and Michel Fessler
Narration written by Frédéric Fougea
Dialogue: Jacques Dubuisson
Director of photography: Martial Barrault
Image synthesis: Mac Guff Ligne. (Producer: Jacques Bled; Design and supervision: Philippe Sonrier)
Music composed by Yvan Cassar
Musical direction: Jean-Marie Leau
Make-up: Maestro Studio F/X
Research: Fabrice Demeter – Collège de France, Paris

Actors: Ashley Abrahams, Jean-Sébastien Allaert, Abigail Andrews, Benjamin Bester, Anzel Biehl, Candice Botha, Jacqui Buchanan, Ashton Bulock, Neil Debruyn, Angelique de Vrye, Smanga Dyasi, Garth Ensley, Tiziana Fenotti, Paul Freeth, Leonhard Freidberg, Joy Gexa, Sue Ellen Goeda, Elize Jacobs, Candice Jafta, Xavier Janson, Mathapelo Kgoleng, Eric Lataste, Lukhanyiso Lubelwana, Brett McLaren, Sam Jack Mabona, Jonathan Maghienda, Jacob Makgoba, Sibusiso Mhlongo, Jone Gayleen Mitchell, Craig Morris, Wright Nugubane, Sharon Gail Opperman, Julisa Petersen, Johannes Ramothope, Kgomotso Rantao, William Rittmann, Syabonga Thati, Danielle Theron, Dominik Uytenbogaardt, Llewellyn van Rooyen, Gavin van Wyk, Alan White, Marion Williams, Zinhle Zuma.

PHOTO CREDITS

All the photos in this book are either from the film *Odyssée de l'espèce* or its making (© *Odyssée de l'espèce*) except:
Pages 10, 14, 16–17, 18, 20–21, 22–23, 24–25, 26–27, 28–29, 30, 32–33, 34(r)–35, 36, 38–39, 40–41, 42–43, 44–45, 46–47, 50, 98, 174–175, 176–177: © Mac Guff Ligne
Pages 15, 34(l): © Imagique-RTBF
Pages 52–53, 54–55, 56, 58(l)–59, 60–61, 64–65, 66–67, 68–69, 70–71, 72–73, 75(t and r), 76–77, 78–79, 80–81, 82–83, 84–85, 100–101, 102–103, 178–179: © Sylvain Legrand/France 3
Page 12(l): © Rudolf König/Jacana
Page 12(r): © Dimijian/PHR/Jacana
Page 13: © Jors and Petra Wener/Jacana
Jacket cover: © Nuit de Chine

ACKNOWLEDGEMENTS

Nicolas Buchet expresses his gratitude to Professor Yves Coppens for the invitation to join the editorial team. Also his sincere thanks to Fabrice Demeter for all his invaluable help and good advice during the course of the work.

Odile Perrard would like to thank in particular Hervé Dresen and Pascal de Cugnac (Transparences Productions 17 juin Productions), Philippe Sonrier (Mac Guff Ligne) and Béatrice Austin (France 3) for their unfailing kindness and helpfulness. Also Nathalie Bailleux and Brigitte Leblanc for their friendly and very productive co-operation.

First published by EPA, an imprint of Hachette-Livre
43 Quai de Grenelle, Paris 75905, Cedex 15, France
Under the title *L'Odysée de l'Espèce*

Senior editor Odile Perrard, with the collaboration of Nathalie Bailleux
Art editor Nancy Dorking
Graphics Nicolas Hubert
Maps Cyrille Suss
Secretarial staff Aurélie Dombes, Christelle Fucili and Laure Sérullaz
Proofreader/Copy editor Marianne Bonneau
Proofreader/Copy editor (science) Fabrice Demeter
Photoengraving APS Chromostyle

English language translation produced by Translate-A-Book, Oxford

This edition published by Hachette Illustrated UK,
Octopus Publishing Group Ltd, 2–4 Heron Quays, London, E14 4JP

ISBN-13 : 978-1-844300-95-2
ISBN-10 : 1-84430-095-1

Printed in Singapore by Tien Wah Press